1st International Conference of Air Quality & Climate Change and 3rd Malaysia Air Quality Annual Symposium

IOP Conference Series: Earth and Environmental Science
Volume 1013

Online
17 – 18 November 2021

ISBN: 978-1-7138-5455-5
ISSN: 1755-1307

Printed from e-media with permission by:

Curran Associates, Inc.
57 Morehouse Lane
Red Hook, NY 12571

Some format issues inherent in the e-media version may also appear in this print version.

Institute of Physics
Dirac House, Temple Back
Bristol BS1 6BE UK

Phone: 44 1 17 929 7481
Fax: 44 1 17 920 0979

techtracking@iop.org

Additional copies of this publication are available from:

Curran Associates, Inc.
57 Morehouse Lane
Red Hook, NY 12571 USA
Phone: 845-758-0400
Fax: 845-758-2633
Email: curran@proceedings.com
Web: www.proceedings.com

TABLE OF CONTENTS

IOP Conf. Series: Earth and Environmental Science 1013 (2022) 011001 doi:10.1088/1755-1315/1013/1/011001

Preface

The 2021 The Air Quality and Climate Change (1st AQCC) and The 3rd Malaysia Air Quality Annual Symposium (MAQAS) organized by Clean Air Forum Society of Malaysia (MyCAS) and supported by Department of Environment Malaysia (Air Division) and Taiwan Association for Aerosol Research (TAAR) was held on the 17th–18th November 2021 in virtual form. Due to the Covid-19 pandemic, ***"Combating Climate Change and Improving Air Quality in a Global Pandemic"*** was chosen as the theme of this year's conference. High challenges have been faced by our team members to manage this event due to the Covid-19. However, with the collaboration and hard work among committee members and participants this event was successfully conducted virtually and managed remotely from Bangi Resort Hotel, Putrajaya and from University Malaysia Terengganu as a second centre. With that, the conference had emphasized interdisciplinary research along with practices and fosters exchanges among scientists, practitioners, professionals and young researchers in finding solutions and sharing latest technology in aerosol and air quality studies and work during this situation. It was an honour to receive more than 50 papers submitted to the conference and I am very pleased to announce that around 17 papers have been selected to be published in the IOP proceeding. All papers which have been accepted were subjected to a rigorous peer-reviewing process to ensure quality and consistency.

Hence, it is vital to tackle air quality issues. The conference also served as an opportunity in establishing networking among scientists, practitioners, professionals and young researchers in international regions in strengthening international collaboration. A wide range of topics was covered from fundamental, applied and technology aspects of aerosol science and air quality that focused on monitoring, control strategies, health impact, policy and management, big data and air pollution modelling using earth observation data, atmospheric transport and emissions from combustion process.

I acknowledge with gratitude the assistance rendered by the scientific team of 2021 AQCC and MAQAS was led by Prof. Dr Juliana Jalaludin and her publication team along-with all the track chairs with continuous support and advice led to the smooth publication of this issue. Last, but not least, we thank the authors, reviewers and the conference secretariat for their contribution which has enabled this publication to become a reality.

Hopefully, all participants and other interested readers will benefit scientifically from the proceedings and also find it stimulating in the process.

With our warmest regards,

Ts Dr Noor Zaitun Yahaya (AHEA)
The Conference Chair

IOP Conf. Series: Earth and Environmental Science 1013 (2022) 011001 doi:10.1088/1755-1315/1013/1/011001

1st International Conference of Air Quality & Climate Change and 3rd Malaysia Air Quality Annual Symposium (1st AQCC & 3rd MAQAS 2021)

MAIN	
Chairman **Deputy of Chairman**	Ts Dr Noor Zaitun binti Yahaya (UMT) Prof. Dr Juliana binti Jalaludin (UPM)
Secretary 1 **Secretary 2** **Secretary 3**	Dr. Mohammad Fakhratul Ridwan bin Zulkifli (UMT) Mrs. Nik Ibtishamiah binti Hj Ibrahim (UM) Dr. Norhidayah binti Abdull (UMP)
Treasurer 1 **Treasurer 2** **Treasurer 3**	Assoc. Prof. Ts. Dr. Nor Haslina Hashim (UTHM) CA Mrs. Hasmidar Adam (Industry) Dr. Mohamed Shahrir bin Mohamed Zahari (UMT)
PORTFOLIO	
Event Manager 1 & 2	Assoc. Prof. Dr. Mohd Sabri Mohd Ghazali (UMT) Assoc. Prof. Sr. Dr. Norhayati Mahyuddin (UM)
Protocol and Special Duties	Mrs. Ezahtulsyareen Ab Rahman (DOE) Mr. Rusdin bin Laiman (UiTM)
Sponsorship and Exhibition	Mrs. Hajah Hartini Haji Mahidin (UiTM) Ir. Dr. Mohd Rehan bin Karim (Industry) Miss Safiah Hani Binti Zulqurnain
Scientifics	Prof. Dr Juliana binti Jalaludin (UPM) Gs. Dr Zulfa Hanan binti Ash`aari (UPM) Dr. Wan Nurdiyana Wan Mansor (UMT) Ts. Nurzawani Binti Md Sofwan (UiTM)
Registration	Dr. Aniza Abu Bakar (UIAM) Miss Safiah Hani Binti Zulqurnain Miss Fahimah Binti Hashim (UMT)
Training and Side Events	Dr. Nurul Ashraf bt. Razali (UMT) Dr. Aniza Abu Bakar (UIAM) Assoc. Prof. Sr. Ts. Dr. Afzan Ahmad Zaini (UNIMAS)
Technical and Media	Mr. Fairus Md. Darus (UiTM) Gs. Ts. Dr. Aliashim Albani (UMT) Ts. Dr. Mohd Arif Rosli (UTHM) Mr. Sharizal Shamsudin (PSAS)

INTERNATIONAL COMMITTEE MEMBERS

Prof. Chuen-Jinn Tsai, *National Yang Ming Chiao Tung University*

Prof. Moo-Been Chang,
National Central University

Prof. Chung-te Lee,
National Central University

Glynda Bathan-Baterina,
Deputy Executive Director of Clean Air Asia (CAA)

Prof. Li-Hao Young,
China Medical University

Prof. Lin-Chi Wang,
Chung Yuan Christian University

The 1st AQCC & 3rd MAQAS 2021 conference was organized by:

The 1st AQCC & 3rd MAQAS 2021conference was sponsored by:

Peer Review Statement

All papers published in this volume have been reviewed through processes administered by the Editors. Reviews were conducted by expert referees to the professional and scientific standards expected of a proceedings journal published by IOP Publishing.

- **Type of peer review:** Double Anonymous
- **Conference submission management system:** Morressier
- **Number of submissions received:** 17
- **Number of submissions sent for review:** 17
- **Number of submissions accepted:** 14
- **Acceptance Rate (Submissions Accepted / Submissions Received × 100):** 82.4
- **Average number of reviews per paper:** 1.99
- **Total number of reviewers involved:** 22
- **Contact person for queries:**
  Name: JULIANA JALALUDIN
  Email: juliana@upm.edu.my
  Affiliation: Universiti Putra Malaysia - Environmental and Occupational Health

IOP Conf. Series: Earth and Environmental Science 1013 (2022) 012001 doi:10.1088/1755-1315/1013/1/012001

A new Android Application Framework (AAF) for API $PM_{2.5}$ data management system

A S Sadun[1], M A Rosli[2*], N H Hashim[2], J A Sukor[1], N Othman[3], N K Cheng[4], T C Si[2] and C Y Ning[2]

[1] Department of Electrical Engineering Technology, Faculty of Engineering Technology, University Tun Hussein Onn Malaysia, Pagoh High Education Hub, KM 1, Jalan Panchor, Muar, Johor, Malaysia
[2] Department of Civil Engineering Technology, Faculty of Engineering Technology, University Tun Hussein Onn Malaysia, Pagoh High Hub, KM 1, Jalan Panchor, Muar, Johor, Malaysia
[3] Faculty of Civil Engineering and Built Environment, University Tun Hussien Onn Malaysia, 86400 Parit Raja, Batu Pahat, Johor, Malaysia
[4] Mybotic, No.8 and 8A, Jalan Universiti 3, Taman Universiti, 86400 Parit Raja, Batu Pahat. Johor, Malaysia
* Correspondence: mohdarif@uthm.edu.my

Abstract. Air Pollution Index (API) is used in Malaysia to measure the ambient air quality and PM2.5 is one of its parameters for ambient air quality monitoring. However, most of the $PM_{2.5}$ samplers available for research and monitoring by universities and monitoring providers are in manual mode in which the data was logged into the equipment data logger and was manually extracted to the personal computer before analysis. Therefore, this project is aimed to propose an Android Application Framework (AAF) containing IoT devices to monitor the real-time ambient air quality and disseminate computed API information through the social media platform. In this study, mobile apps which are Blynk was used and enabled the user to control remotely desirable outcomes at any time. E-sampler IoT interface, mobile application, data management, data management system, API judgment and information dissemination are proposed to make AAF succeed. As a result, the proposed AAF has been fully utilized in obtaining real-time pollutants data and finally analyzing $PM_{2.5}$ to obtain real-time API information. IoT device has successfully interface with E-Sampler, to obtain raw data and be able to be downloaded for further analysis. In conclusion, AAF has functioned great and information was disseminated to the public through social media.

1. Introduction

Rapid urbanization and industrialization, aggressive economic activities and excessive exploitation are shifting the natural beauty environment towards degradation. As a developing country, Malaysia is currently facing a series of challenges from environmental issues. Climate change, global warming, depletion of the ozone layer, various environmental pollution, ecosystem degradation, and excretion of harmful waste are a few anthropogenic catastrophes [1]. The Rakyat Post (TRP) online news reported that according to the World Air Quality Index (WAQI), Malaysia has an average API reading of 266 declared as the most polluted country in the world in 2019 due to the occurrence of haze [2]. In Malaysia, Air Pollution Index (API) is used as the ambient air quality measurement. The World Health Organization (WHO) new Malaysia ambient air quality guidelines (AQG) in Table 1 are designed to offer guidance in reducing the health impacts of air pollution[3]. API calculation in Table 2 is based on

Air Quality Index (AQI) that has been accepted at the international level by the United States Environmental Protection Agency (USEPA) [4]. The air pollutants concentration limit has been strengthened in stages until 2020 with three interim targets were set for all API parameters as a guideline [5].

Table 1. New Malaysia Ambient Air Quality Standard [5].

Pollutants	Averaging Time	Ambient Air Quality Standard (μg/m³) IT-1 (2015)	IT-2 (2018)	Standard (2020)
Particulate Matter with the size	1 Year	50	45	40
of less than 10 microns (PM_{10})	24 Hour	150	120	100
Particulate Matter with the size	1 Year	35	25	15
of less than 10 microns ($PM_{2.5}$)	24 Hour	75	50	35
Sulfur Dioxide (SO_2)	1 Hour	350	300	250
	24 Hour	105	90	80
Nitrogen Dioxide (NO_2)	1 Hour	320	300	280
	24 Hour	75	75	70
Ground Level Ozone (O_3)	1 Hour	200	200	180
	8 Hour	120	120	100
*Carbon Monoxide	1 Hour	35	35	30
	8 Hour	10	10	10

* mg/m³

API is developed in a way easily understood ranges of values as a means of reporting the quality of air instead of the actual concentration of air pollutants. A sub-index formula for $PM_{2.5}$ will be implemented to the raw data. The ranges also reflect its effects on human health ranging from good to hazardous. The API ranges can be classified into five ranges, which are good (0-50), moderate (51-100), unhealthy (101-200), very unhealthy (201-300), and hazardous (greater than 300) [6]. The color indication in Table 3 will be used in this project to show the real-time API status at any time.

Table 2. Sub-index formula for $PM_{2.5}$ [6].

API	Breakpoint of concentration X=PM2.5 concentration (24h average, unit;μg/m³)	Equation for API
0-50	$0 \le X \le 12.0$	API=4.1667 x X
51-100	$12.1 \le X \le 75.5$	API=0.7741 x (X-12.1) + 51
101-200	$75.5 \le X \le 150.4$	API=1.3218 x (X-75.5) + 101
201-300	$150.5 \le X \le 250.4$	API=0.9909 x (X-150.5) + 201
301-400	$250.4 \le X \le 350.4$	API=0.9909 x (X-250.5) + 301
401-500	$350.4 \le X \le 500.4$	API=0.6604 x (X-350.5) + 401

Table 3. Reading for API.

Reading for API	
0-50	Good
51-100	Moderate
101-200	Unhealthy
201-300	Very Unhealthy
Above 300	Hazardous

Met-One E-Sampler instrument is used currently in UTHM Parit Raja to measure particulate. Several sensors were used to obtain raw data. This included Wind Speed and Wind Direction sensor, Ambient Relative Humidity (RH) sensor, Internal Relative Humidity (RH) sensor, Ambient, Temperature (AT) sensor, Pressure sensor, and Flow sensor [7]. In the conventional framework shown in Figure 1, the E-Sampler instrument is set up to collect data manually from the equipment data logger and send it to the laptop or PC with the installed software. Then, the raw data is downloaded in Excel acceptable format, imported to Excel, and proceeded with further analysis.

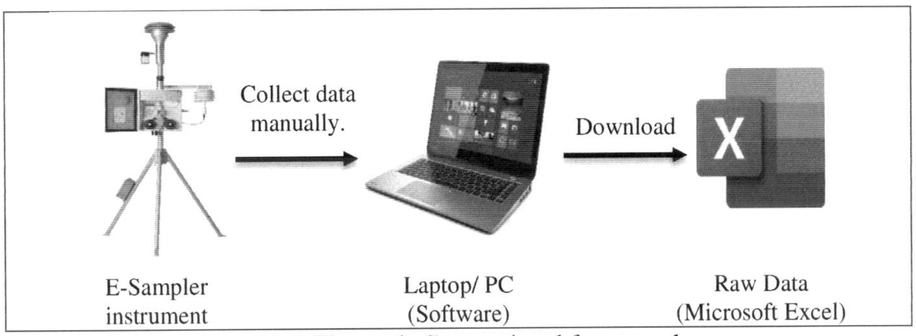

Figure 1. Conventional framework.

UTHM Parit Raja campus has encountered this challenge where lacking an appropriate and advanced framework to monitor the real-time ambient air quality. The current collecting data process is using the conventional framework that required more labor effort and consumed time. PM2.5 sampler available for research and monitoring by university and monitoring provider is in manual mode in which the data was logged into the equipment data logger and it was manually extracted from the data logger to personal computer to analyze. The problem become worst during the Movement Control Order (MCO) 2020 as well as during the haze event or rainy season as the access to the site for data extraction was limited. Real-time data collection becomes impractical and impossible. This is then delayed the data analysis process as well as incapable to provide real-time API information to the public. To accomplish advanced real-time monitoring on ambient air, an appropriate android application framework (AAF) in Figure 2 is required with the integration of hardware, IoT device, software, and application. This framework included six stages: The E-Sampler, Mobile Apps, Data Management, Data Management System, API Judgement, and Information Dissemination. This makes the process from collecting data to distributing information will become simpler, faster, and easier.

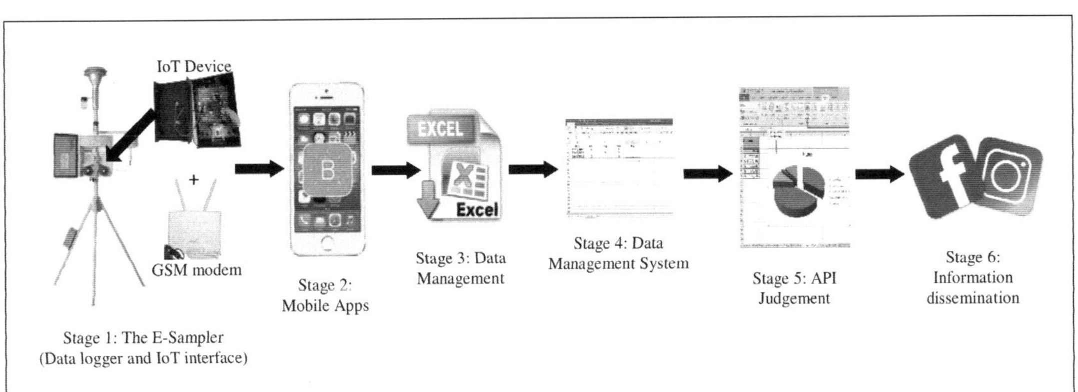

Figure 2: Android Application Framework (AAF).

1.1. Objectives
This study aims to propose a framework containing an IoT device to monitor the real-time ambient air quality and disseminate computed API information through the social media platform. The objectives of this study are:

1. To compute the Air Pollution Index (API) of previous $PM_{2.5}$ collected data for March, April, June, and July 2020 using a data management system.
2. To establish a medium for computed information dissemination through social media Facebook and Instagram.
3. To propose a new android application framework containing IoT devices to monitor real-time ambient air quality for a selected model of E-sampler.

1.2. Project Scope and Limitation
The scope of this study centers on the ability of an android application framework to monitor the real-time ambient air quality in the UTHM Parit Raja campus area with the aids of calculation of API and distribute ambient air information to the public. The study will revolve around workability and functionality on the new proposed android application framework (AAF) toward data collection and data analyses to obtain real-time API information. Some limitations of the study are:

1. The study will be conducted outdoor only, indoor air quality is not examined. Due to the MCO period during the COVID-19 pandemic, the data management system was computed using the previously collected $PM_{2.5}$ data.
2. The study is focusing on API parameter $PM_{2.5}$ only.
3. The social media platform users are limited to Facebook and Instagram.
4. Stage 1 to stage 3 of the proposed AAF is at a preliminary stage. There is no existing framework able to collect data continuously.

2. Methodology
To monitor real-time ambient air quality in the UTHM Parit Raja campus area, an advance proposed android application framework (AAF) that included an E-Sampler instrument, IoT, mobile apps, data management system and social media platform was proposed. A selected outdoor location in the UTHM Parit Raja campus is fixed for placing the E-Sampler instrument as well as the IoT device and GSM modem. Previous real-time $PM_{2.5}$ data were collected and used for analysis purposes in the formulated Microsoft Excel, which is named as a data management system. In this project, Excel is fully used for the calculation of API and analysis of ambient air quality status. Lastly, the analyzed API result and API-related information are scattered and disseminated through social media platforms, Facebook, and Instagram.

The methods applied in this chapter are based on the study objectives as mentioned, which are then to be achieved. This study focuses on the proposed android application framework which included six stages. The methodology flow chart of this study is shown in Figure 3.

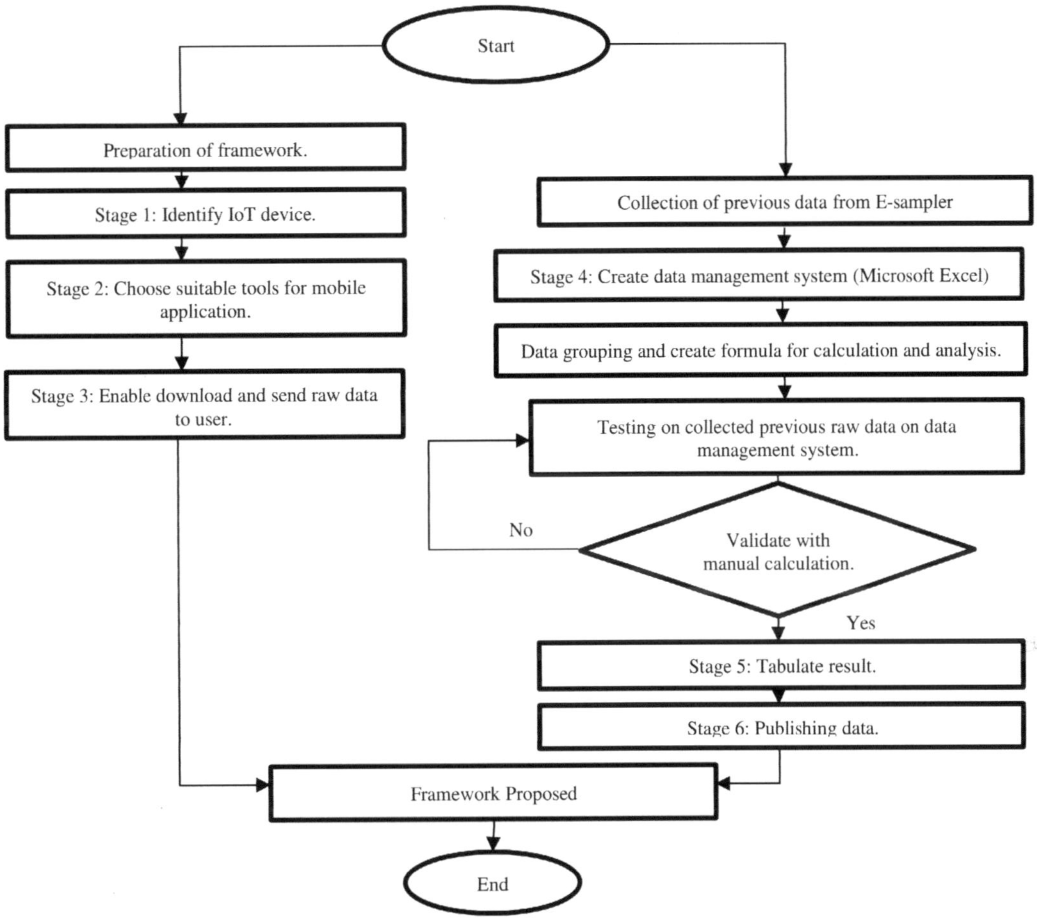

Figure 3. Methodology flow chart of this study.

2.1. Stage 1 – The E-Sampler (Data Logger and IoT Interface)

Met-One E-Sampler instrument was selected for the usage of this project. It was ensured that it has the memory of 4,369 internal data loggers that record data overtime for 182 days or 1 record/hour in 3 days or 1 record/min, and in relation to having built-in sensors and external sensors. A suitable IoT device was proposed to make the connection between E-Sampler the mobile apps.

2.2. Stage 2- Mobile application

A suitable tool (Blynk app) was selected in this stage. The Blynk app in Figure 4 was downloaded and was set up by creating a new Blynk account as well as creating a new project. Blynk was connected to the hardware through supported cellular, and the initial empty project was then modified by adding widgets. Blynk app was designed with buttons to control E-Sampler remotely, display sensors data, stored data, and display plotted graphs. Blynk app was designed to present periodic data, which included live or real-time, 15 minutes, 30 minutes, an hour, a day, a week, or a month. The displayed data was designed, color was used to differentiate data and improve visual satisfaction.

IOP Conf. Series: Earth and Environmental Science 1013 (2022) 012001 doi:10.1088/1755-1315/1013/1/012001

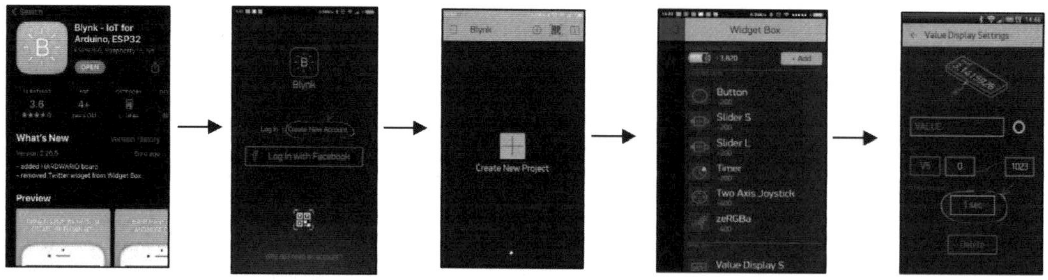

Figure 4. Blynk app setting up process.

2.3. Stage 3- Data management

As mentioned previously, the designed Blynk app with buttons and settings was able to control E-Sampler remotely, display sensors data, stored data, and display plotted graphs. Generated and stored data was requested to download and export. The exported data file format was supported by Microsoft Excel.

2.4. Stage 4- Data management system

At this stage, Microsoft Excel was used for the data management system. Appropriate formulas and equations were designed for the overall process of obtaining the final API information based on the USEPA standard. Moreover, the detailed conditioned formula was constructed to obtain secondary data that will be used for analyzing purposes. That included the formula for obtaining $PM_{2.5}$ concentration tier, API, API tier, and API color. The raw data were then grouped and organized according to time, week, year, month, day and tabulated raw data into breakdown data.

2.5. Stage 5- API Judgment

The coloring method is used to determine API status for a certain period. The color indicates ambient air status according to USEPA standard, either good (blue), moderate (green), unhealthy (yellow), very unhealthy (orange), hazardous (red), or emergency (red). A conditioned formula has been set to generate a tabulated result of API with the color. Another method used was plotted graph method. The daily plot graph visualizes better with the changes of ambient air quality based on hourly measurement. Instead of that, the daily graph of API was also plotted monthly as well.

2.6. Stage 6- Information dissemination

There are several approaches available for data dissemination on API-related information. From the choices of website, short message service (SMS), e-mail, social media, radio, and television, in this project, the proposed approach is via social media platforms, Facebook, and Instagram. API information will be updated on these social media from time to time. At this stage, a poster template was designed using Canva apps. In the poster template, several API-related information was designed. This included the university and faculty logo and the sources of posted API information by FTK UTHM were inserted. Besides, API status with its color indication as well as the color description was designed in the poster.

3. Result and discussion

The proposed android application framework (AAF) in this project involved the usage of hardware and software. E-Sampler instrument is connected to IoT device and Blynk app through a cellular connection.

The data management system has been designed to perform calculations and analyses on previously collected raw data. Social media platform had successfully created and finally, API information is updated to the platform, Facebook and Instagram periodically.

3.1. Location of data collection

UTHM Parit Raja campus is located at Batu Pahat, Johor. It was selected to collect $PM_{2.5}$ from ambient air to be analyzed to obtain API-related information on the location. Through observation from Google Map as Figure 5, the UTHM Parit Raja campus was surrounded by plenty of plantation and pervious surface, compared to buildings and houses. The northern part of the campus is a huge lawn, the southern part has rows of houses along the road, shop lot and buildings were built compactly along the main road from east to west direction.

Figure 5. UTHM-Parit Raja Campus (Google Map).

3.2. Stage 1- The E-Sampler (Data Logger and IoT Interface)

IoT device and E-Sampler data logger were connected physically in a laboratory to test the functionality of the IoT device. Global System for Mobile Communication (GSM) modem is a wireless modem that communicates with the GSM network, to generate, transmit and decode data from IoT devices to Blynk apps. A block diagram is shown in Figure 6 on the connection between the E-Sampler instrument, IoT device and GSM modem. The proposed system at this stage is still undergoing development and improvement and is yet to be fully established and validated.

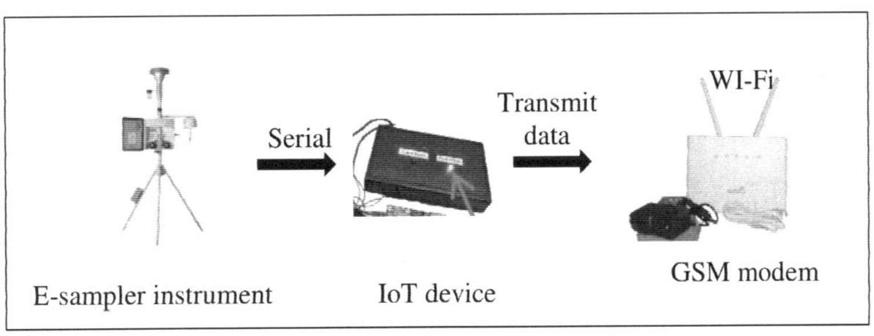

Figure 6. Block diagram of hardware.

The proposed IoT device was tested out in the laboratory together with an E-Sampler data logger as shown in Figure 7 below. Figure 8 shows on the IoT device, data status is lighted indicate that data is successfully transmitted from E-Sampler data logger to Blynk apps.

Figure 7. Testing on the functionality of the IoT device.

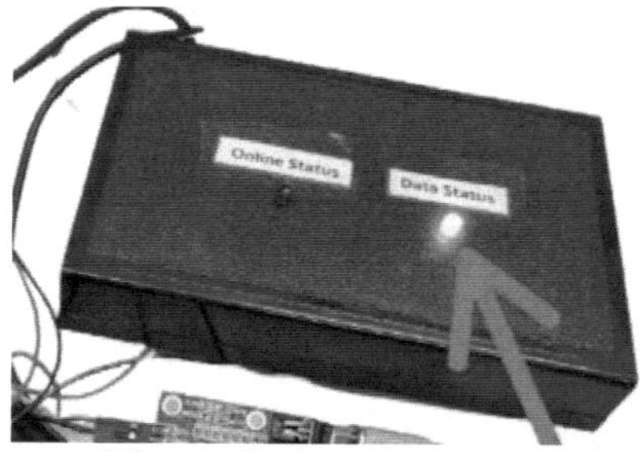

Figure 8. Data transmitting.

3.3. Stage 2- Mobile Application

As a preliminary result, the Blynk app is currently able to achieve all the functions smoothly. Data were collected, stored, and displayed as a graph. Referring to Figure 9, sensors data and plotted graph was displayed with desirable time frame.

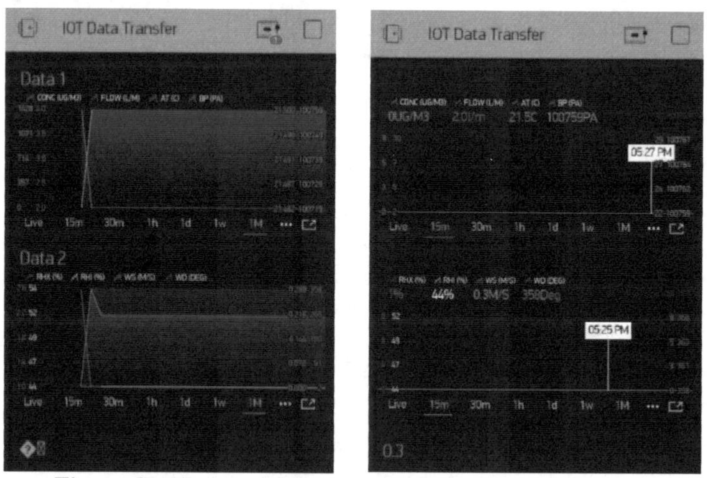

Figure 9. Displayed data information on the Blynk app.

In the plotted graph, the color presented indicates specific raw data. These eight raw data were prescribed with their specific color and this shows in Table 4.

Table 4. Color indication on raw data.

Data	Colour	Abbreviation	Raw Data
1	Green	CONC (UG/M3)	$PM_{2.5}$ concentration (μg/m^3)
	Purple	FLOW (L/M)	Volumetric flow rate (l/m)
	Blue	AT (C)	Ambient temperature (°C)
	Orange	BP (PA)	Barometric pressure (pa)
2	Red	RHX (%)	External relative humidity (%)
	White	RHI (%)	Internal relative humidity (%)
	Blue	WS (M/S)	Wind speed (m/s)
	Purple	WD (DEG)	Wind direction (°)

3.4. Stage 3 - Data Management
As a preliminary result, the download function in Blynk apps works successfully. Data were successfully downloaded as shown in Figure 10 in CSV file format that was supported by Microsoft Excel. Besides, selected raw data are also currently able to send to the user through an email address. Nevertheless, continuous observation of the function of the Blynk app guarantees the accuracy of the data collected.

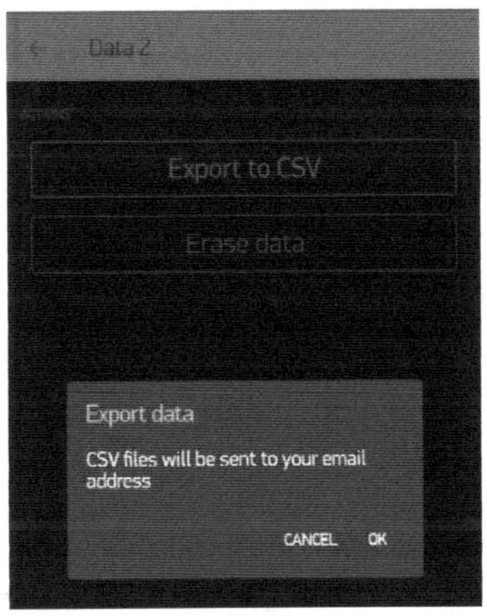

Figure 10. Download display in Blynk app.

3.5. Stage 4 - Data Management System
The raw data used in this system is collected manually in previous. There is two excel file in total to generate API information for the month of March ~ April and the month of June ~ July. In each Excel, there are six sheets to store raw data, group and break down data, perform calculations, tabulate data and finally, generate complete API information. In "Raw Data" sheets, raw data has successfully been imported into the system without amendment and editing on it. This is shown in Figure 11. The formula needed for API calculation is pasted into the raw data sheet and acts as the guideline for designing the formula in the following steps.

Raw Data Set 1

	Time	Conc (µG/	Flow (l/m)	WS (M/S)	WD (Deg)
5	3/4/2020 11:00	5	1.8	0.9	112
6	3/4/2020 12:00	5	2	1.1	132
7	3/4/2020 13:00	5	2	1	121
8	3/4/2020 14:00	5	2	1.2	198
9	3/4/2020 15:00	7	2	0.8	120
10	3/4/2020 16:00	8	2	1.6	304
11	3/4/2020 17:00	11	2	1.4	353
12	3/4/2020 18:00	16	2	1.5	126
13	3/4/2020 19:00	22	2	0.8	31
14	3/4/2020 20:00	24	2	0.3	115
15	3/4/2020 21:00	26	2	0.4	122
16	3/4/2020 22:00	35	2	0.3	117
17	3/4/2020 23:00	35	2	0.3	124
18	3/5/2020 0:00	17	2	0.3	125
19	3/5/2020 1:00	20	2	0.3	125
20	3/5/2020 2:00	20	2	0.3	123
21	3/5/2020 3:00	15	2	0.3	118
22	3/5/2020 4:00	15	2	0.3	117
23	3/5/2020 5:00	13	2	0.3	119
24	3/5/2020 6:00	14	2	0.3	127
25	3/5/2020 7:00	14	2	0.3	119
26	3/5/2020 8:00	13	2	0.3	115
27	3/5/2020 9:00	11	1.9	0.3	115
28	3/5/2020 10:00	10	2	0.6	106
29	3/5/2020 11:00	10	2	0.7	104
30	3/5/2020 12:00	9	2	0.6	107
31	3/5/2020 13:00	11	2	0.5	122
32	3/5/2020 14:00	10	2	0.7	123
33	3/5/2020 15:00	11	2	0.6	121

Calculation Formula Set 1

	API	BREAKPOINT OF CONCENTRATION	EQUATION FOR API
	X=PM2.5 concentration(24h average, unit;µg/m3)		
GOOD	0-50	0≤X≤12.0	API=4.1667 x X
MODERATE	51-100	12.1≤X≤75.5	API=0.7741 x (X-12.1) + 51
UNHEALTHY	101-200	75.5≤X≤150.4	API=1.3218 x (X-75.5) + 101
VERY UNHEALTHY	201-300	150.5≤X≤250.4	API=0.9909 x (X-150.5) + 201
VERY UNHEALTHY	301-400	250.4≤X≤350.4	API=0.9909 x (X-250.5) + 301
HAZARDOUS	401-500	350.4≤X≤500.4	API=0.6604 x (X-350.5) + 401

	API	BREAKPOINT OF CONCENTRATION	EQUATION FOR API
	X=PM10 concentration(24h average, unit;µg/m3)		
GOOD	0-50	X≤50	API=X
MODERATE	51-100	50≤X≤150	API=50 + [(X-50) x 0.5]
UNHEALTHY	101-200	150≤X≤350	API=100 + [(X-50) x 0.5]
VERY UNHEALTHY	201-300	350≤X≤420	API=200 + [(X-350) x 1.4286]
VERY UNHEALTHY	301-400	420≤X≤500	API=300 + [(X-420) x 1.25]
HAZARDOUS	401-500	X>500	API=400 + (X-500)

Remarks
1. Refer Cell SFS334 for amendment made on data set.

User Manual | Raw Data | Clean Data | Helper | Statistic | API_PM_x

Figure 11. Raw data set.

Pivot table of computed data from raw data shown in Figure 12 when raw data was successfully organized and grouped into time, week, month, day, and hour.

Copy & Paste Data Set

	Time	Conc (µG/M3)	Flow (l/m)	WS (M/S)	WD (Deg)
5	3/4/2020 11:00	5	1.8	0.9	112
6	3/4/2020 12:00	5	2	1.1	132
7	3/4/2020 13:00	5	2	1	121
8	3/4/2020 14:00	5	2	1.2	198
9	3/4/2020 15:00	7	2	0.8	120
10	3/4/2020 16:00	8	2	1.6	304
11	3/4/2020 17:00	11	2	1.4	353
12	3/4/2020 18:00	16	2	1.5	126
13	3/4/2020 19:00	22	2	0.8	31
14	3/4/2020 20:00	24	2	0.3	115
15	3/4/2020 21:00	26	2	0.4	122
16	3/4/2020 22:00	35	2	0.3	117
17	3/4/2020 23:00	35	2	0.3	124
18	3/5/2020 0:00	17	2	0.3	125
19	3/5/2020 1:00	20	2	0.3	125
20	3/5/2020 2:00	20	2	0.3	123
21	3/5/2020 3:00	15	2	0.3	118
22	3/5/2020 4:00	15	2	0.3	117
23	3/5/2020 5:00	13	2	0.3	119
24	3/5/2020 6:00	14	2	0.3	127
25	3/5/2020 7:00	14	2	0.3	119
26	3/5/2020 8:00	13	2	0.3	115
27	3/5/2020 9:00	11	1.9	0.3	115

Breakdown Date & Timestamp Details

Time	week	year	month	day	hour	Conc (µ	Flow (l/	WS (M/	WD (De
3/4/2020 11:00	10	2020	3	4	11	5	1.8	0.9	112
3/4/2020 12:00	10	2020	3	4	12	5	2	1.1	132
3/4/2020 13:00	10	2020	3	4	13	5	2	1	121
3/4/2020 14:00	10	2020	3	4	14	5	2	1.2	198
3/4/2020 15:00	10	2020	3	4	15	7	2	0.8	120
3/4/2020 16:00	10	2020	3	4	16	8	2	1.6	304
3/4/2020 17:00	10	2020	3	4	17	11	2	1.4	353
3/4/2020 18:00	10	2020	3	4	18	16	2	1.5	126
3/4/2020 19:00	10	2020	3	4	19	22	2	0.8	31
3/4/2020 20:00	10	2020	3	4	20	24	2	0.3	115
3/4/2020 21:00	10	2020	3	4	21	26	2	0.4	122
3/4/2020 22:00	10	2020	3	4	22	35	2	0.3	117
3/4/2020 23:00	10	2020	3	4	23	35	2	0.3	124
3/5/2020 0:00	10	2020	3	5	0	17	2	0.3	125
3/5/2020 1:00	10	2020	3	5	1	20	2	0.3	125
3/5/2020 2:00	10	2020	3	5	2	20	2	0.3	123
3/5/2020 3:00	10	2020	3	5	3	15	2	0.3	118
3/5/2020 4:00	10	2020	3	5	4	15	2	0.3	117
3/5/2020 5:00	10	2020	3	5	5	13	2	0.3	119
3/5/2020 6:00	10	2020	3	5	6	14	2	0.3	127
3/5/2020 7:00	10	2020	3	5	7	14	2	0.3	119
3/5/2020 8:00	10	2020	3	5	8	13	2	0.3	115
3/5/2020 9:00	10	2020	3	5	9	11	1.9	0.3	115

Figure 12. Computed data from raw data.

Secondary data were obtained when detailed conditioned formulas and equations were designed in the “Helper” sheet. Secondary data included concentration tier, API, API tier, and API color information. This is shown in Figure 13 below. Lastly, the calculated API results are successfully tabulated in the “Statistic” sheet that shows in Figure 14 below.

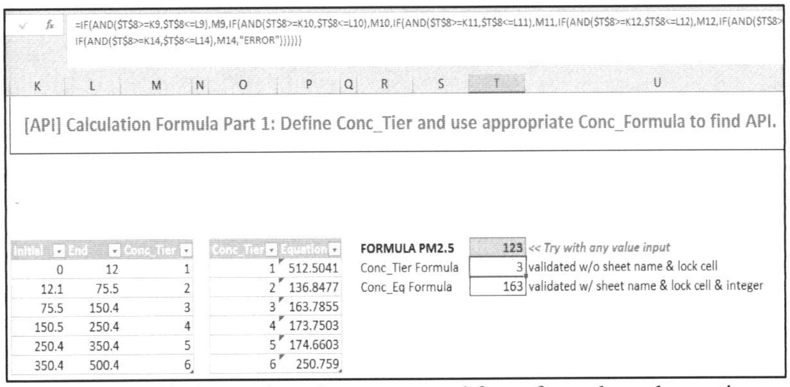

Figure 13. Secondary data generated from formula and equation.

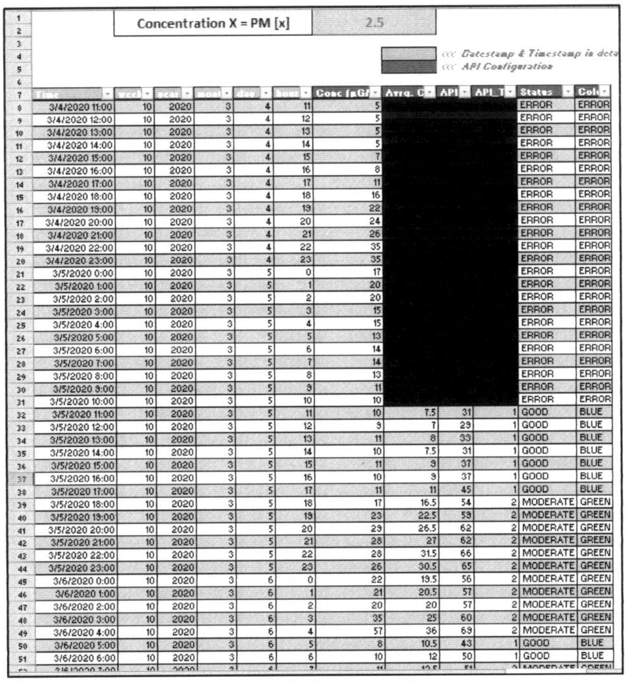

	Time	week	year	mont	day	hour	Conc [µG/	Avrg. C	API	API_T	Status	Col
8	3/4/2020 11:00	10	2020	3	4	11	5	[redacted]			ERROR	ERROR
9	3/4/2020 12:00	10	2020	3	4	12	5				ERROR	ERROR
10	3/4/2020 13:00	10	2020	3	4	13	5				ERROR	ERROR
11	3/4/2020 14:00	10	2020	3	4	14	5				ERROR	ERROR
12	3/4/2020 15:00	10	2020	3	4	15	7				ERROR	ERROR
13	3/4/2020 16:00	10	2020	3	4	16	8				ERROR	ERROR
14	3/4/2020 17:00	10	2020	3	4	17	11				ERROR	ERROR
15	3/4/2020 18:00	10	2020	3	4	18	16				ERROR	ERROR
16	3/4/2020 19:00	10	2020	3	4	19	22				ERROR	ERROR
17	3/4/2020 20:00	10	2020	3	4	20	24				ERROR	ERROR
18	3/4/2020 21:00	10	2020	3	4	21	26				ERROR	ERROR
19	3/4/2020 22:00	10	2020	3	4	22	35				ERROR	ERROR
20	3/4/2020 23:00	10	2020	3	4	23	35				ERROR	ERROR
21	3/5/2020 0:00	10	2020	3	5	0	17				ERROR	ERROR
22	3/5/2020 1:00	10	2020	3	5	1	20				ERROR	ERROR
23	3/5/2020 2:00	10	2020	3	5	2	20				ERROR	ERROR
24	3/5/2020 3:00	10	2020	3	5	3	15				ERROR	ERROR
25	3/5/2020 4:00	10	2020	3	5	4	15				ERROR	ERROR
26	3/5/2020 5:00	10	2020	3	5	5	13				ERROR	ERROR
27	3/5/2020 6:00	10	2020	3	5	6	14				ERROR	ERROR
28	3/5/2020 7:00	10	2020	3	5	7	14				ERROR	ERROR
29	3/5/2020 8:00	10	2020	3	5	8	13				ERROR	ERROR
30	3/5/2020 9:00	10	2020	3	5	9	11				ERROR	ERROR
31	3/5/2020 10:00	10	2020	3	5	10	10				ERROR	ERROR
32	3/5/2020 11:00	10	2020	3	5	11	10	7.5	31	1	GOOD	BLUE
33	3/5/2020 12:00	10	2020	3	5	12	9	7	29	1	GOOD	BLUE
34	3/5/2020 13:00	10	2020	3	5	13	11	8	33	1	GOOD	BLUE
35	3/5/2020 14:00	10	2020	3	5	14	10	7.5	31	1	GOOD	BLUE
36	3/5/2020 15:00	10	2020	3	5	15	11	9	37	1	GOOD	BLUE
37	3/5/2020 16:00	10	2020	3	5	16	10	9	37	1	GOOD	BLUE
38	3/5/2020 17:00	10	2020	3	5	17	11	11	45	1	GOOD	BLUE
39	3/5/2020 18:00	10	2020	3	5	18	17	16.5	54	2	MODERATE	GREEN
40	3/5/2020 19:00	10	2020	3	5	19	23	22.5	59	2	MODERATE	GREEN
41	3/5/2020 20:00	10	2020	3	5	20	29	26.5	62	2	MODERATE	GREEN
42	3/5/2020 21:00	10	2020	3	5	21	28	27	62	2	MODERATE	GREEN
43	3/5/2020 22:00	10	2020	3	5	22	28	31.5	66	2	MODERATE	GREEN
44	3/5/2020 23:00	10	2020	3	5	23	26	30.5	65	2	MODERATE	GREEN
45	3/6/2020 0:00	10	2020	3	6	0	22	19.5	56	2	MODERATE	GREEN
46	3/6/2020 1:00	10	2020	3	6	1	21	20.5	57	2	MODERATE	GREEN
47	3/6/2020 2:00	10	2020	3	6	2	20	20	57	2	MODERATE	GREEN
48	3/6/2020 3:00	10	2020	3	6	3	35	25	60	2	MODERATE	GREEN
49	3/6/2020 4:00	10	2020	3	6	4	57	36	69	2	MODERATE	GREEN
50	3/6/2020 5:00	10	2020	3	6	5	8	10.5	43	1	GOOD	BLUE
51	3/6/2020 6:00	10	2020	3	6	6	10	12	50	1	GOOD	BLUE

Figure 14. Tabulated result.

3.6. Stage 5 - API Judgement

Figure 15 shows the generated March and April API information is tabulated according to time and was highlighted with color. From the results obtained in Figure15(a), there are 8 slots highlighted yellow, which indicated the ambient air was at an unhealthy state. Exception of 10th March, 11th March and 12th March having moderate status, majority results show that from early 10 am to evening 6 pm, the ambient air status was good. In general, from 6 pm onward until the next morning at 10 am, the ambient air is at a moderate state. On the other hand, the ambient air has exceeded the 2020 ambient air quality limit standard even though the result shows it was at a moderate state. According to Figure 15(b), from 8 pm to 10 am, there is a random day when ambient air exceeded the limit standard. However, on 10th March, at 2 am ambient air quality had achieved the highest API value (110.5µg/m^3).

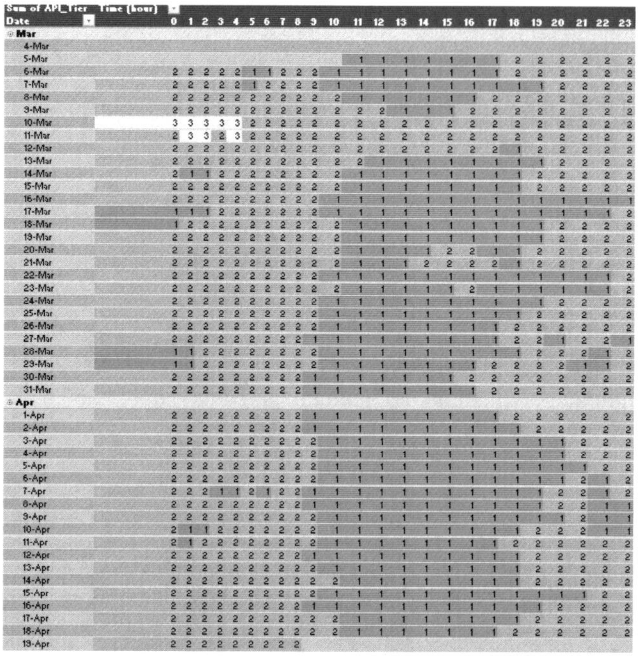

(a)

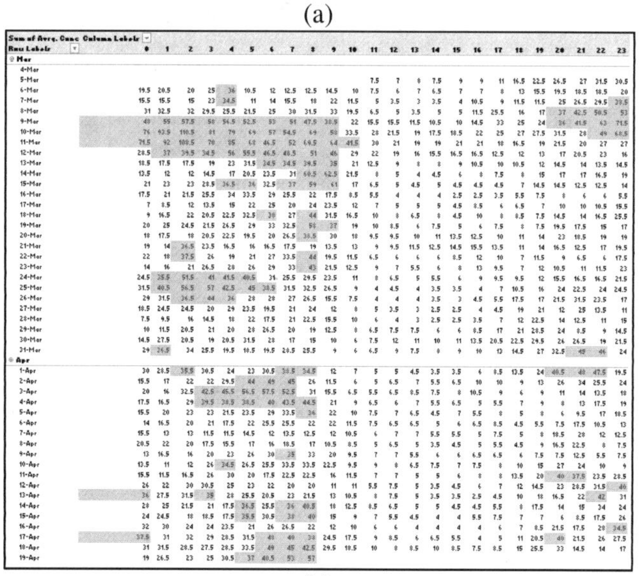

(b)

Figure 15. (a) API color indication result from 04th March 2020 until 19th April 2020; (b) API from 04th March 2020 until 19th April 2020.

Figure 16 shows the generated June and July API information is tabulated according to time and was highlighted with color. The result obtained in Figure 16(a) shows that the majority of API results from 10 am to 8 pm were highlighted blue which represents ambient air was in a good state. However, minor results show highlighted green from 8 pm to 10 am, this represents the ambient air was at a moderate state during this period. Besides, the ambient air has exceeded the 2020 ambient air quality limit standard even though the result shows it was at a moderate state. According to Figure 16(b), there are 6 slots highlighted red that indicate the over-limit time and concentration value. Nevertheless, they were exceeding only 1 and 2 μg/m^3 on the random day mostly in the morning.

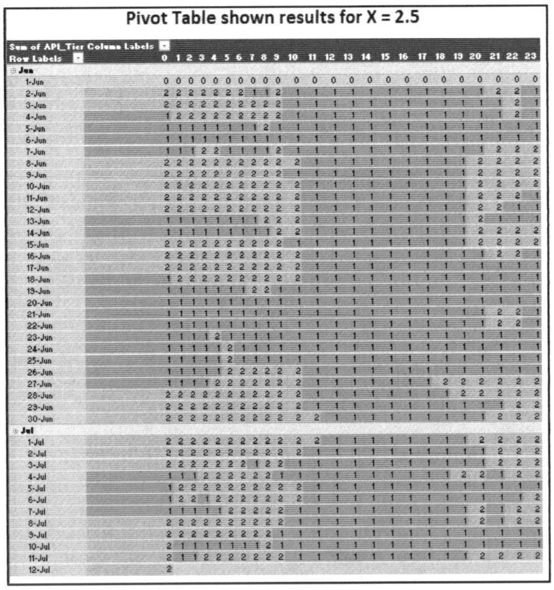

Pivot Table shown results for X = 2.5

Sum of API_Tier Column Labels / Row Labels	0	1	2	3	4	5	6	7	8	9	10	11	12	13	14	15	16	17	18	19	20	21	22	23
Jun																								
1-Jun	0	0	0	0	0	0	0	0	0	0	0	0	0	0	0	0	0	0	0	0	0	0	0	0
2-Jun	2	2	2	2	2	2	2	1	1	2	1	1	1	1	1	1	1	1	1	1	1	2	2	1
3-Jun	2	2	2	2	2	2	2	2	2	2	1	1	1	1	1	1	1	1	1	1	1	1	2	1
4-Jun	1	2	2	2	2	2	2	2	2	2	1	1	1	1	1	1	1	1	1	1	1	1	2	1
5-Jun	1	1	1	1	1	1	1	1	2	1	1	1	1	1	1	1	1	1	1	1	1	1	1	1
6-Jun	1	1	1	1	1	1	1	1	1	1	1	1	1	1	1	1	1	1	1	1	1	1	1	1
7-Jun	1	1	1	2	2	1	1	1	1	2	1	1	1	1	1	1	1	1	1	1	1	2	2	2
8-Jun	2	2	2	2	2	2	2	2	2	2	2	1	1	1	1	1	1	1	1	1	2	2	2	2
9-Jun	2	2	2	2	2	2	2	2	2	2	1	1	1	1	1	1	1	1	1	1	2	2	2	2
10-Jun	2	2	2	2	2	2	2	2	2	2	2	1	1	1	1	1	1	1	1	1	2	2	2	2
11-Jun	2	2	2	2	2	2	2	2	2	2	2	1	1	1	1	1	1	1	1	1	2	2	2	1
12-Jun	2	2	2	2	2	2	2	2	2	2	2	1	1	1	1	1	1	1	1	1	2	2	1	1
13-Jun	1	1	1	1	1	1	1	1	2	2	2	1	1	1	1	1	1	1	1	1	2	1	1	1
14-Jun	1	1	1	1	1	1	1	1	1	2	2	1	1	1	1	1	1	1	1	1	2	2	2	2
15-Jun	2	2	2	2	2	2	2	2	2	2	1	1	1	1	1	1	1	1	1	1	2	2	2	2
16-Jun	2	2	2	2	2	2	2	2	2	2	2	1	1	1	1	1	1	1	1	1	1	2	2	1
17-Jun	2	2	2	2	2	2	2	2	2	2	2	1	1	1	1	1	1	1	1	1	1	1	1	1
18-Jun	1	2	2	2	2	2	2	2	2	2	2	1	1	1	1	1	1	1	1	1	1	1	1	1
19-Jun	1	1	1	1	1	1	1	2	2	1	1	1	1	1	1	1	1	1	1	1	1	1	1	1
20-Jun	1	1	1	1	1	1	1	1	1	1	1	1	1	1	1	1	1	1	1	1	1	1	1	1
21-Jun	1	1	1	1	1	1	1	1	1	1	1	1	1	1	1	1	1	1	1	1	1	2	2	1
22-Jun	1	1	1	1	1	1	1	1	1	1	1	1	1	1	1	1	1	1	1	1	1	2	2	1
23-Jun	1	1	1	1	2	1	1	1	1	1	1	1	1	1	1	1	1	1	1	1	1	1	1	1
24-Jun	1	1	1	1	1	2	1	1	1	1	1	1	1	1	1	1	1	1	1	1	1	1	1	1
25-Jun	1	1	1	1	1	2	1	1	1	1	1	1	1	1	1	1	1	1	1	1	1	1	1	1
26-Jun	1	1	1	1	1	2	2	2	2	2	2	1	1	1	1	1	1	1	1	1	1	1	1	1
27-Jun	1	1	1	1	2	2	2	2	2	2	2	1	1	1	1	1	1	1	2	2	2	2	2	2
28-Jun	2	2	2	2	2	2	2	2	2	2	2	1	1	1	1	1	1	1	1	2	2	2	2	2
29-Jun	2	2	2	2	2	2	2	2	2	2	2	1	1	1	1	1	1	1	1	1	1	1	2	2
30-Jun	2	2	2	2	2	2	2	2	2	2	2	2	1	1	1	1	1	1	1	1	1	2	2	2
Jul																								
1-Jul	2	2	2	2	2	2	2	2	2	2	2	2	1	1	1	1	1	1	1	1	2	2	2	2
2-Jul	2	2	2	2	2	2	2	2	2	2	2	1	1	1	1	1	1	1	1	1	1	2	2	2
3-Jul	2	2	2	2	2	2	2	1	2	2	1	1	1	1	1	1	1	1	1	1	1	2	2	2
4-Jul	1	1	1	2	2	2	2	2	2	1	1	1	1	1	1	1	1	1	1	2	2	1	2	2
5-Jul	1	2	2	2	2	2	2	2	2	2	2	1	1	1	1	1	1	1	1	1	1	1	1	1
6-Jul	1	2	2	1	2	2	2	2	2	2	2	1	1	1	1	1	1	1	1	1	1	1	1	2
7-Jul	1	1	1	1	1	2	2	2	2	2	1	1	1	1	1	1	1	1	1	1	2	1	2	2
8-Jul	2	2	2	2	2	2	2	2	2	2	1	1	1	1	1	1	1	1	1	1	2	1	2	2
9-Jul	2	2	2	2	2	2	2	2	2	1	1	1	1	1	1	1	1	1	1	1	1	1	1	1
10-Jul	2	1	1	1	1	1	1	1	2	1	1	1	1	1	1	1	1	1	1	1	1	1	1	1
11-Jul	2	1	1	2	2	2	2	2	2	2	1	1	1	1	1	1	1	1	1	1	2	2	2	2
12-Jul	2																							

(a)

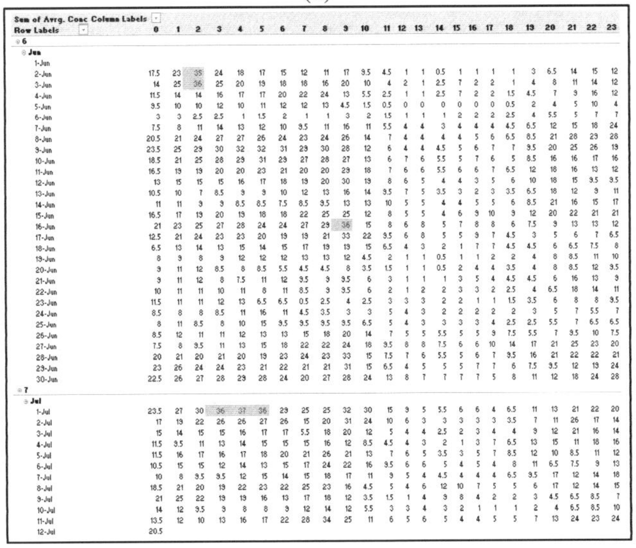

Sum of Avrg. Conc Column Labels / Row Labels	0	1	2	3	4	5	6	7	8	9	10	11	12	13	14	15	16	17	18	19	20	21	22	23
6																								
Jun																								
1-Jun																								
2-Jun	17.5	23	35	24	18	17	15	12	11	17	9.5	4.5	1	1	0.5	1	1	1	1	3	6.5	14	15	12
3-Jun	14	25	36	25	20	19	18	18	16	20	10	4	2	1	2.5	7	2	2	1	4	8	11	14	12
4-Jun	11.5	14	14	16	17	17	20	22	24	13	5.5	2.5	1	1	2.5	7	2	2	1.5	4.5	7	9	16	12
5-Jun	9.5	10	10	12	10	11	12	12	13	4.5	1.5	0.5	0	0	0	0	0	0	0.5	2	4	5	10	4
6-Jun	3	3	2.5	2.5	1	1.5	2	1	1	3	2	1.5	1	1	1	2	2	2	2.5	4	5.5	5	7	7
7-Jun	7.5	8	11	14	13	12	10	9.5	11	16	11	5.5	4	4	3	4	4	4	4.5	6.5	12	15	18	24
8-Jun	20.5	21	24	27	27	26	24	23	24	26	14	7	4	4	4	4	5	6	6.5	8.5	21	28	29	28
9-Jun	23.5	25	29	30	32	32	31	29	30	28	12	6	4	4	4.5	5	6	7	7	9.5	20	25	26	19
10-Jun	18.5	21	25	28	29	31	29	27	28	27	13	6	7	6	5.5	5	7	6	5	8.5	16	16	17	16
11-Jun	16.5	19	19	20	20	23	21	20	20	29	18	7	6	6	5.5	6	6	7	6.5	12	18	16	13	12
12-Jun	13	15	15	15	16	17	18	19	20	30	19	8	6	5	4	4	3	5	6	10	18	15	9.5	9.5
13-Jun	10.5	10	7	8.5	9	9	10	12	13	16	14	9.5	7	5	3.5	3	2	3	3.5	6.5	18	12	9	11
14-Jun	11	11	9	9	8.5	8.5	7.5	8.5	9.5	13	13	10	5	5	4	4	5	5	6	8.5	21	16	15	17
15-Jun	16.5	17	19	20	19	18	18	22	25	25	12	8	5	5	4	6	9	10	9	12	20	22	21	21
16-Jun	21	23	25	27	28	24	24	27	29	36	15	8	6	8	5	7	8	8	6	7.5	9	13	13	12
17-Jun	12.5	21	24	23	23	20	19	19	21	33	22	9.5	6	8	5	5	9	7	4.5	3	5	6	7	6.5
18-Jun	6.5	13	14	13	15	14	15	17	19	19	15	6.5	4	3	2	1	7	7	4.5	4.5	6	6.5	7.5	8
19-Jun	8	9	8	9	12	12	12	13	13	12	4.5	2	1	1	0.5	1	1	2	2	4	8	8.5	11	10
20-Jun	9	11	12	8.5	8	8.5	5.5	4.5	4.5	8	3.5	1.5	1	1	0.5	2	4	4	3.5	4	8	8.5	12	9.5
21-Jun	9	11	12	8	7.5	11	12	9.5	9	9.5	6	3	1	1	1	3	5	4	4.5	4.5	6	16	13	9
22-Jun	10	11	11	10	11	8	11	8.5	9	9.5	6	2	1	2	2	3	3	2	2.5	4	6.5	18	14	11
23-Jun	11.5	11	11	12	13	6.5	6.5	0.5	2.5	4	2.5	3	3	3	2	2	1	1	1.5	3.5	6	8	8	9.5
24-Jun	8.5	8	8	8.5	11	16	11	4.5	3.5	3	3	5	4	3	2	2	2	2	2	3	5	7	5.5	7
25-Jun	8	11	8.5	8	10	15	9.5	9.5	9.5	9.5	6.5	5	4	3	3	3	3	4	2.5	2.5	5.5	7	6.5	6.5
26-Jun	8.5	12	11	11	12	13	13	15	18	20	14	7	5	5	5.5	5	5	9	7.5	5.5	7	9.5	10	7.5
27-Jun	7.5	8	9.5	11	13	15	18	22	22	24	18	9.5	8	8	7.5	6	6	10	14	17	21	25	23	20
28-Jun	20	21	20	21	20	19	23	24	23	33	15	7.5	7	6	5.5	5	6	7	9.5	16	21	22	22	21
29-Jun	23	26	24	24	23	21	22	21	21	31	15	6.5	4	5	5	5	7	7	6	7.5	9.5	12	19	24
30-Jun	22.5	26	27	28	29	28	24	20	27	28	24	13	8	7	7	7	7	5	8	11	12	18	24	28
7																								
Jul																								
1-Jul	23.5	27	30	36	37	36	29	25	25	32	30	15	9	5	5.5	6	6	4	6.5	11	13	21	22	20
2-Jul	17	19	22	26	26	27	26	15	20	31	24	10	6	3	3	3	3	3	3.5	7	11	26	17	14
3-Jul	15	14	15	15	16	17	17	5.5	18	20	12	5	4	4	2.5	2	3	4	4	9	12	21	16	14
4-Jul	11.5	9.5	11	13	14	15	15	15	16	12	8.5	4.5	4	3	2	1	3	7	6.5	13	15	11	18	16
5-Jul	11.5	16	17	16	17	18	20	21	26	21	13	7	6	5	3.5	3	5	7	8.5	12	10	8.5	11	12
6-Jul	10.5	15	15	12	14	13	15	17	24	22	16	9.5	6	6	5	4	5	4	8	11	6.5	7.5	9	13
7-Jul	10	8	9.5	9.5	12	15	14	15	18	17	11	9	5	4	4.5	4	4	4	6.5	9.5	17	12	14	18
8-Jul	18.5	21	20	19	22	23	22	25	23	16	4.5	5	4	6	12	10	7	5	5	6	17	12	14	15
9-Jul	21	25	22	19	19	16	13	17	18	12	3.5	1.5	1	4	9	8	4	2	2	3	4.5	6.5	8.5	7
10-Jul	14	12	9.5	9	8	8	9	12	14	12	5.5	3	3	4	3	2	1	1	1	2	4	6.5	8.5	10
11-Jul	13.5	12	10	13	16	17	22	28	34	25	11	6	5	6	5	4	4	5	5	7	13	24	23	24
12-Jul	20.5																							

(b)

Figure 16. (a) API color indication result from 01st June 2020 until 12th July 2020; (b) API from 01st June 2020 until 12th July 2020.

3.7. Stage 6 – Information Dissemination

As a result, social media platform has slowly engaged with the public. Figure 17(a) shows the post engagement has increased 480%, and to people reached has increased 1033% at the Facebook platform. For the Instagram account, Figure 17(b) shows the total current engagement is with 85 followers. This shows that social media platform is successful in engaging the public and information are disseminated easily.

(a) (b)

Figure 17. (a) Facebook platform engagement; (b) Instagram platform engagement.

The designed poster shown in Figure 18 has been posted on both Facebook and Instagram platforms. In the poster, API-related information such as API status and description, API range, and real-time API were published.

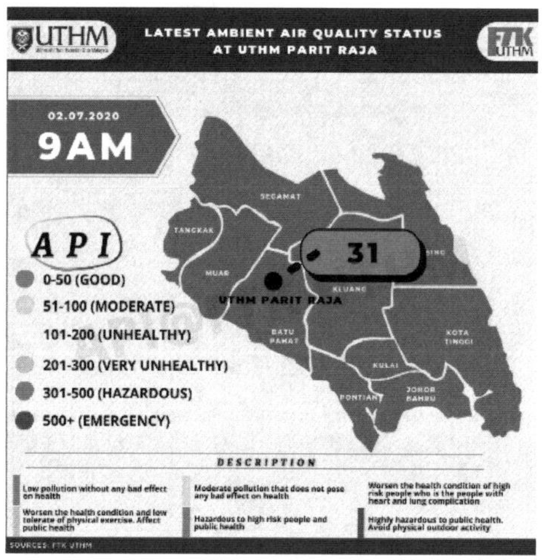

Figure 18. Poster API.

4. Conclusion

The android application framework has functioned great in this whole project. Although stage 1 to stage 3 is still under development, however, the operator can obtain real-time information from AAF developing process. The implementation of IoT devices into the framework has made the whole process from collecting data until obtaining raw data fast and simple. This also illustrates the importance of AAF in making life easier, mainly when the covid-19 pandemic hits Malaysia. With the aid of AAF, operators can obtain raw data downloaded from the Blynk app and be able to monitor the ambient air quality anytime at anywhere.

According to the results from the previous chapter, it can be observed that in the month of March and April 2020, ambient air quality is poorer compared with the month of June and July 2020. The amount of concentration exceeding the standard limit in June and July is 6, with the highest concentration of 37 $\mu g/m^3$. This shows the ambient air in UTHM Parit Raja in these two months has improved. Instead, the highest concentration from the month of March and April is 110.5$\mu g/m^3$ (ambient air condition: unhealthy). This will worsen the health condition of high-risk people who is with heart and lung complications. The ambient air condition also increases aggravation of heart and lung disease and premature mortality in persons with cardiopulmonary disease and the elderly. Sensitive groups are encouraged to avoid prolonged exertion and stay indoors. The changes in ambient air may closely relate to the activities held on campus during the period. As several standard operating procedures (SOP) were implemented at the UTHM Parit Raja campus from April 2020, attendance of workers, lecturers and students were decreased. Hence, transportation and laboratory work on campus was greatly reduced. Therefore, this is believing that ambient air quality is greater in June and July.

Three objectives of the project were achieved throughout the whole process of study. At the end of this study, API for the month of March, April, June, and July 2020 were calculated and computed. The proposed new data management system is fully utilized in getting API in UTHM Parit Raja in this project. Besides, Facebook page and Instagram account were established to disseminate API-related information. By spreading the project account to the public, the response from people is good. Friends and strangers started to follow and get engaged with the Facebook and Instagram platforms to get API information. This is good to increase public awareness on the importance of air quality, as it will largely affect human daily activities. Lastly, a new android application framework is proposed to monitor real-time ambient air quality in the UTHM Parit Raja campus area. This new framework is useful, especially since the user can control remotely on the E-Sampler to collect desired data.

4.1. Recommendations

This study had served to improve in collecting data from manually collecting data to using IoT devices and software to collect real-time data and shorten the analysis process by using a data management system. However, further improvement and modification can still be made for future studies of similar nature. A few recommendations are given as such:

1. Improve the functionality and efficiency of the data management system. The current system is designed just for $PM_{2.5}$. This limits the usage of the system to calculate and analyze other API parameters. The system is operated manually, this may consume time when a large amount of database is imported into the system.
2. Study on the extent, the impact of the new framework to the university and public. This framework may lead UTHM Parit Raja to become an API hotspot to provide real-time API to the public.
3. Social media platforms should be shared with more friends and relatives, to continuously increase the engagement of the platform with the public.

5. References

[1] Mei N S, Wai C W, and Ahamad R, 2016, "Environmental Awareness and Behaviour Index for Malaysia," *Procedia - Soc. Behav. Sci.*, vol. 222, no. 07, pp. 668–675,

[2] Nazari T, 2019 ,"Haze : M ' sia declared # 1 most air polluted country in the world According to the World Air Quality Index (WAQI), Malaysia recorded an average API reading of 266 on," *TRP*, Sep. 19.

[3] World Health Organization (WHO), 2005,"WHO Air Quality Guidelines for particulate matter, ozone, nitrogen dioxide, and sulfur dioxide," *Summ. risk Assess.*, vol. 51, no. 6, pp. 565–573.

[4] Ministry of E. Department of Environment, "Official Portal of Department of Environment - Air Pollutant Index (API)," 2020. https://www.doe.gov.my/portalv1/en/info-umum/english-air-pollutant-index-api/100.

[5] M. A. A. Q. DOE, "New Malaysia Ambient Air Quality Standard," p. 1989, 2020, [Online]. Available: http://www.doe.gov.my/portalv1/wp-content/uploads/2013/01/Air-Quality-Standard-BI.pdf.

[6] Department of Environmental, 2019, "API Calculation".

[7] Met One Instrument, 2011,"E-SAMPLER-9800 MANUAL Rev M".

[8] New Straits Times, 2019, "Haze crisis : Still no breather for much of Malaysia," no. 108, Kuala Lumpur, pp. 1–5, Sep. 22.

Acknowledgments
The authors would like to thank the special work by Farrah Nuramanina binti Mohd Anuar from the Faculty of Civil Engineering and Built Environment, for her data collection and support.

Effects of Payload on Non-exhaust PM Emissions from Hybrid Electric Vehicle during Braking Sequences

E Wirojsakunchai[1], C Plengsa-Ard[1] and W Songkitti[1]*

[1] Faculty of Engineering, Kasetsart University, 50 Ngam Wong Wan Rd, Lat Yao Chatuchak Bangkok 10900
*Corresponding author

Abstract. Vehicles equipped with internal combustion engines are known as one of the most important Particulate Matters (PM) emissions sources. Many countries are aware of this issue and interested in employing more electric vehicles to reduce this emission. However, various past research claims that electric vehicles also emitted PM as conventional vehicles due to their non-exhaust emissions such as brake wear, tyre wear, road surface wear, and road dust resuspension. In addition, substantial evidence showed that there was indeed a positive correlation between the weight of vehicles and amounts of non-exhaust PM emissions. This study aims to measure on-road non-exhaust PM emissions from a hybrid electric vehicle during braking sequences. An onboard PM measuring device is attached to the side of the tested hybrid electric vehicle. PM measurements are monitored during the braking sequence in the electrified vehicle mode. Studies of increasing payloads that might affect the tendency of non-exhaust PM emissions are observed. The PM emission pattern during the braking sequence is captured by the current PM measuring setup as seen in the literature. The braking pattern (hard vs. soft brake) shows distinct amounts in PM emissions by a factor of two. Based on experimental data, it is found out that the additional payload of approximately 70 kg increases the amount of non-exhaust PM emissions by almost 20%.

1. Introduction

Particulate Matter (PM) is one of the world's most problematic pollutants in terms of harm to human health, resulting in the air quality index. PM is often divided into PM10 and PM2.5, representing particles with a diameter of less than 10 μm and 2.5 μm, respectively. The primary source of PM can be found in urban areas, especially under traffic congestion. One of the strategies being adopted in many countries to improve the air quality index is incentivizing the electrification of passenger cars. The switch to Electric Vehicle (EV) fleets has been proposed to solve air pollution, offer zero emissions, and promise cleaner air for everyone. However, when modelling the impact of EVs on air quality [1-2], it was found out that fleet electrification would not significantly reduce PM emissions due to the importance of non-exhaust PM emissions. Conventionally, vehicles emit PM through their exhaust and non-exhaust sources, namely, tyre wear, brake wear, road surface wear, and resuspension of road dust [3]. These non-exhaust emissions tend to contain mostly PM10, but a significant proportion of the emissions contains fine PM2.5 as well. The chemical characteristics of non-exhaust PM emissions vary per source but are mainly made up of heavy metals such as zinc (Zn), copper (Cu), iron (Fe) and lead (Pb), among others [4]. The road abrasion and tyre wear are caused by the friction between the tyre tread and road surface. Brake wear is caused by the friction between the brake pads and the wheel. Finally, road dust resuspension is caused by the diffusion of air current underneath and behind vehicles [5].

Among these non-exhaust PM emissions, brake wear is considered as a significant contribution because of the high frequency of its usage. It was already hypothesized that PM emissions from brake wear were hig.hly influenced by vehicle weight [6-7]. Several individual studies measuring this non-exhaust emission differentiate between passenger cars and light-duty vehicles (LDVs). Despite varying definitions for the weight of vehicle categories, the explicit result was that LDVs emitted more non-exhaust PM from brake wear than passenger cars [8]. It was also mentioned that the inertia weight, while the vehicle was being stopped, was one of the most critical factors contributing to brake wear rates. However, no test has been done to confirm this observation yet.

Therefore, this research focuses on investigating of non-exhaust PM emissions emitted from a hybrid electric vehicle during braking sequences. The test is done by using a PM Mobile onboard measuring equipment. This mobile on-board measurement is done by attaching directly onto a moving vehicle nearby a wheel. Payloads of vehicles are varied to study the effects of weights on non-exhaust PM emissions during braking sequences.

2. Materials

An experimental setup is done by attaching real-time PM monitoring equipment at the left front brake on a hybrid midsize passenger car wheel, as shown in Figure 1. The specification of the tested vehicle is shown in Table1. All hardware of the current real-time PM monitoring equipment is shown in Figure 2. Table 2 shows the specification of the dust sensor used in the current study.

Figure 1. Measurement setup with instruments

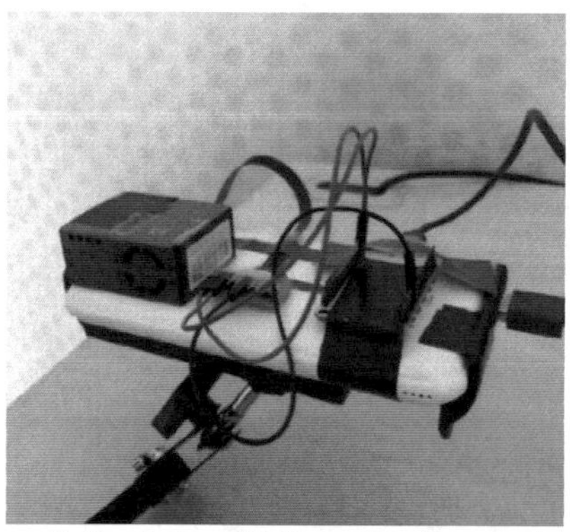

Figure 2. Real-time PM monitoring equipment

Table 1. Specifications of Tested Vehicle

Parameter	Index
Engine	
Type	4 Cylinder in-line, SOHC 16 valve
Capacity (cc.)	1,798
Bore x Stroke (mm.)	80.5 x 88.3
Compression Ratio	13:1
Max. Output EEC net kw (ps)/rpm	72 (98) / 5,200
Max. Torque EEC net Nm (kg-m)/rpm	142(14.5)/3600
Electric Motor	
Type	Synchronous Motor with Permanent Magnet
Max. Voltage (V.)	600
Max. Output (kw.)	53
Max. Torque	163
Hybrid Battery	
Type	Nickel-metal Hydride
Voltage (V.)	201.6
No. of Module	28 Modules 168 Cells
Capacity (Amr-Hr)	6.5(3)
Engine and Electric Motor Max. Output kw(ps)	90(122)

Table 2. Specification of dust sensor PMS5003

Parameter	Index	Unit
Range of measurement	0.3~1.0 1.0~2.5 2.5~10	Micrometer (μm)
Counting Efficiency	50%@0.3μm 98%@>=0.5μm	
Effective Range	0~500	$\mu g/m^3$
Maximum Range	≥1000	$\mu g/m^3$
Resolution	1	$\mu g/m^3$
Maximum Consistency Error	±10%@100~500 $\mu g/m^3$ ±10 $\mu g/m^3$@0~100 $\mu g/m^3$	
Standard Volume	0.1	Litre (L)
Single Response Time	<1	Second (s)
Total Response Time	≤10	Second (s)
DC Power Supply	Typ:5.0 Min:4.5 Max:5.5	Volt (V)
Active Current	≤100	Milliampere (mA)
Standby Current	≤200	Microampere (μA)
Interface Level	L<0.8@3.3H>2.7@3.3	Volt (V)
Working Temperature Range	-10~+60	°C
Working Humidity Range	0~99%	
Storage Temperature Range	-40~+80	°C
MTTF	≥3	Year (Y)
Physical Size	50×38×21	Millimeter (mm)

3. Methodology

The measurement concept is as follows. PM is detected by a dust sensor connecting to the ESP32 (WIFI+Blutooth) board run by Arduino IDE for sending and receiving commands. ESP32 needs to upload a code program, namely, Bluetooth32 to connect a Bluetooth, PMS_MCU to EMS32 for command sensor, and Plantower PMS5003 for detecting PM1, PM2.5, and PM10. A schematic of all equipment connections is shown in Figure 3.

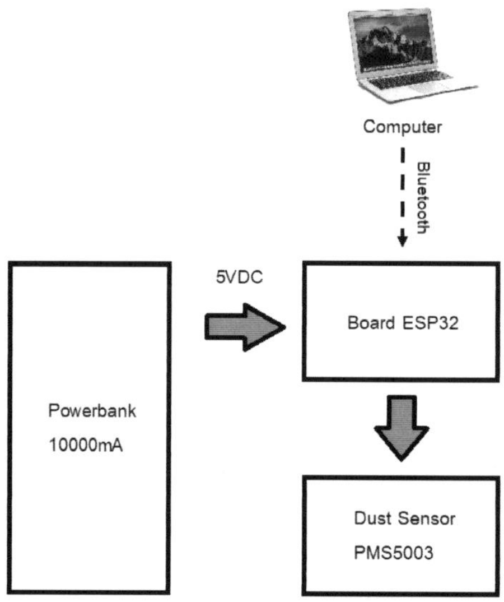

Figure 3. A schematic of the experimental setup

PM readings from the current setup are compared to PM standard measuring tool [9] as seen in Figure 4. Results show good agreements between two PM measuring devices.

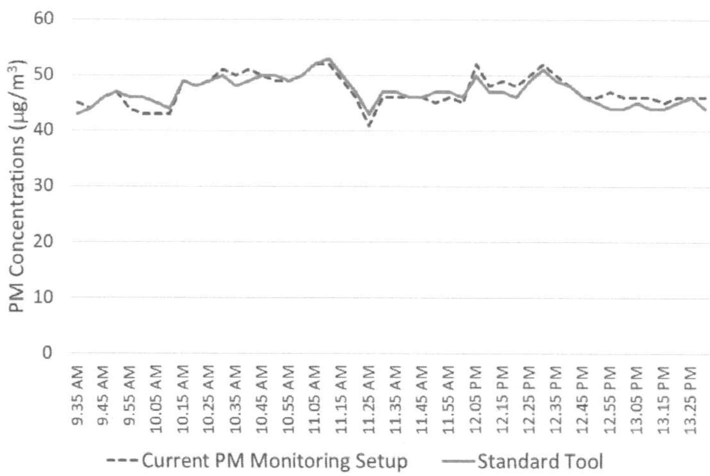

Figure 4. A comparison of PM concentrations measuring by the current setup and a PM standard tool

In the current study, all tests are performed on a hybrid electric vehicle. The braking and resuspension systems are Original Equipment Manufacturer (OEM) and the vehicle is always in routine regular maintenance. All tests are done on the same road within a village to minimize PM diffusion from another vehicle. The vehicle velocities are increased from 0 to 40 km/h to ensure that the vehicle is in the electrified mode and all PM that are emitted from the tested vehicle is non-exhaust. The stopping distance from the velocity of 40 km/h is set within 5 meters until the vehicle is fully stopped. Tests are repeated by increasing various payloads on the tested vehicle. The payload is increased by adding approximately 70 kg, ranging from 2-7 passengers in the vehicle. The analysis of relationships between non-exhaust PM emissions and payloads are shown in the next section.

4. Result and discussion

Figure 5 shows an example of raw data of PM measurement during a break sequence. At 0 seconds, the onboard PM measuring device is started while the vehicle is in park mode. PM emissions are read as the background level (approximately 22 μg/m^3). At 10 sec, the vehicle is started, and the vehicle speed is increased. PM readings during this period are due to resuspension of road dust effect (approximately 33 μg/m^3). The brake is applied until the speed reaches 40 km/h (approximately at 20 seconds). PM emissions increase rapidly. The vehicle is completely stopped at 26 seconds. However, PM emissions continuously increase for 5 seconds and slowly decrease until PM readings are equal to the background level. Note that the vehicle is always in the electrified mode during this sequence, as shown in Figure 5. The behavior of non-exhaust PM emissions that are currently measured corresponds to trends observed in past works of literature [9].

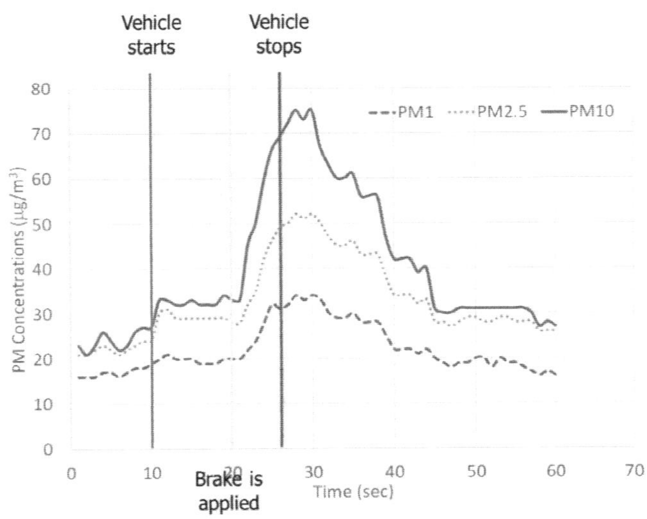

Figure 5. An example of non-exhaust PM measurement during a braking sequence

Prior to variation of payload tests, the effect of braking behavior is on trial. Figures 6 and 7 show a comparison between soft (slowly decrease the vehicle velocity) and complex (rapidly decrease the vehicle velocity) brake tests. As we see from both figures, the hard brake generates more PM emissions than the soft one. This corresponds to results demonstrated in, for example, ref. [11]. However, both methods of testing yield similar PM emissions trends. For the current study, the hard brake test is chosen for investigating the payload effect.

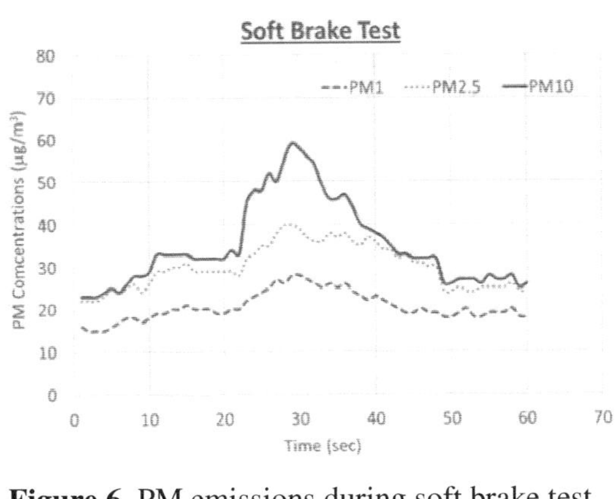

Figure 6. PM emissions during soft brake test

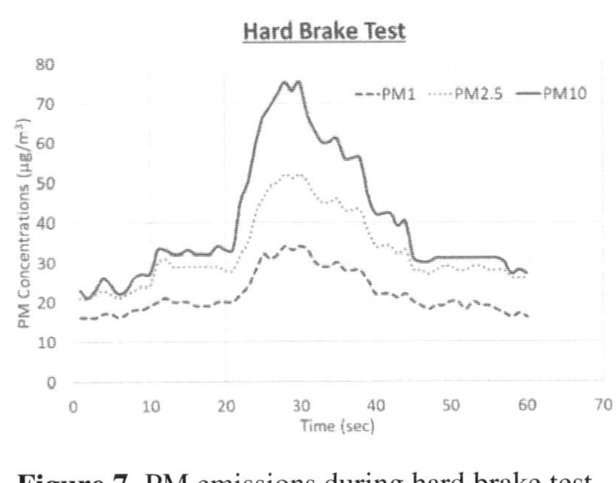

Figure 7. PM emissions during hard brake test

Past literature indicated that the moisture level of the road surface might affect the retention of dust on the road [12]. Experiments on PM measurements on various temperatures and humidity are done. An example of PM2.5 measuring data is shown in Table 3. The first column indicates time, temperature, and humidity levels. The second column shows the average values of PM measuring data between the vehicle starts and when the brake is applied. The third column represents the maximum value of data set when data are measured. The fourth column shows the difference between the third and second columns representing the range of PM emissions. Based on these results, there is no significant impact of moisture level on non-exhaust PM measurement found in the current study.

Table 3. Measuring data of PM2.5 on various moisture levels.

PM2.5	**Normal Drive** *(μg/m³)*	**Peak** *(μg/m³)*	**Ranges of PM Emissions** *(μg/m³)*
Time: 10.20 AM Temperature: 29 °C Humidity: 70%	55	75	+20
Time: 01.20 PM Temperature: 34 °C Humidity: 54%	56	76	+20
Time: 04.30 PM Temperature: 33 °C Humidity: 60%	52	70	+18

Figure 8 represents the value of a range of PM emissions as described in the previous paragraph on various payloads. Five tests are performed by increasing the payload at approximately 60-70 kg each. Considering two passengers are the baseline case (+0 kg), results show that with higher payload on the vehicle, non-exhaust PM including PM1, PM2.5, and PM10 emits greater during braking sequence. Each passenger can yield almost up to a 20% increase in ranges of PM emissions. With six passengers in the vehicle compared to two passengers, non-exhaust PM can emit more than two times. An example plot is made to demonstrate the linear relationship between the payload and the value of a range of PM2.5 emissions. Results are shown in Figure 9. Apart from PM2.5, the plots of PM1 and PM10 are identical to that shown in Figure 9.

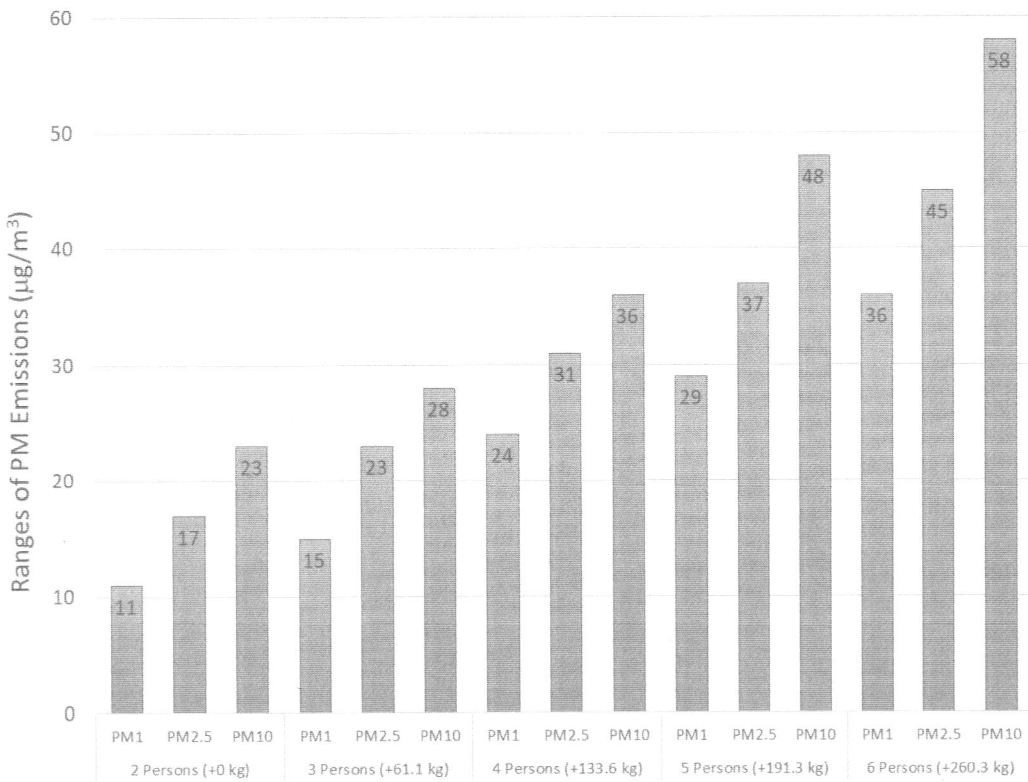

Figure 8. Effects of payload on non-exhaust PM emissions

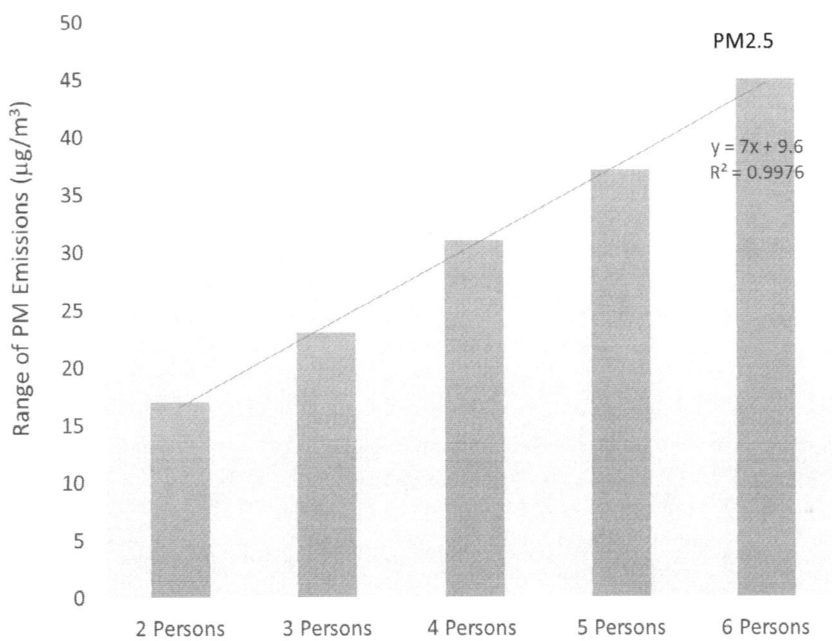

Figure 9. Effects of payload on non-exhaust PM emissions

5. Conclusions

This research study focuses on the effect of payloads on non-exhaust PM emissions from the hybrid electric vehicle. The onboard PM monitoring device is attached nearby the wheel on the vehicle for real-time PM measurement. Payload is varied from two to six passengers and data are collected during braking sequences. Results show that by increasing the payload at approximately 60-70 kg for each test, PM emissions can be increased up to 20%. With six passengers in the vehicle compared to two passengers, non-exhaust PM can emit more than 2 times. A linear relationship can also be found between PM emissions and increased payloads.

Acknowledgement

This research study is financially supported by the Faculty of Engineering, Kasetsart University via Graduate Research Scholarship Contract No. 62/09/ME/D.ENG. Thanks to Thana Jongcharoensiri, Phathit Sungthong, and Vitsarut Namsiriyothin for data collection and the tested vehicle.

References

[1] Soret, A., Guevara, M., & Baldasano, J. M. (2014). The potential impacts of electric vehicles on air quality in the urban areas of Barcelona and Madrid (Spain). Atmospheric Environment, 99, 51-63. doi:https://doi.org/10.1016/j.atmosenv.2014.09.

[2] Kuenen, J. J. P., Visschedijk, A. J. H., Jozwicka, M., & Denier van der Gon, H. A. C. (2014). TNO-MACC_II emission inventory; a multi-year (2003–2009) consistent high-resolution European emission inventory for air quality modelling. Atmos. Chem. Phys., 14(20), 10963-10976. doi:10.5194/acp-14- 10963-2014

[3] Timmers, V. R. J. H. and Archten, P. A. J. (2016). Non-exhaust PM Emissions from Electric Vehicles. Atmospheric Environments. v134 pp. 10-17.

[4] Thorpe, A., & Harrison, R. M. (2008). Sources and properties of non-exhaust particulate matter from road traffic: A review. Science of The Total Environment, 400(1), 270-282. doi:https://doi.org/10.1016/j.scitotenv.2008.06.

[5] Simons, A. J. T. I. J. o. L. C. A. V. (2013). Road transport: new life cycle inventories for fossil-fuelled passenger cars and non-exhaust emissions in Ecoinvent v3. (9), 1299-1313.

[6] Barlow, T. (2014). Briefing Paper on Non-exhaust Particulate Emissions from Road Transport

[7] Garg, B., Cadle, S., Mulawa, P., & Groblicki, P. (2000). Brake Wear Particulate Matter Emissions. Environmental Science & Technology - ENVIRON SCI TECHNOL, 34. doi:10.1021/es001108h

[8] Luekewille, A., Bertok, I., Amann, M., Cofala, J., Gyarfas, F., Heyes, C., . . . Schoepp, W. (2001). A Framework to Estimate the Potential and Costs for the Control of Fine Particulate Emissions in Europe. Retrieved from IIASA, Laxenburg, Austria: http://pure.iiasa.ac.at/id/eprint/6497/

[9] Online air quality report Kasetsart university.: https://airq.ku.ac.th/ (accessed on 29 April 2020).

[10] Mathissen, M., Grochowicz, J., Schmidt, C., Vogt, R., Farwick zum Hagen, F. H., Grabiec, T., Grigoratos, T. (2018). A novel real-world braking cycle for studying brake wear particle emissions. Wear, 414-415, 219-226.

[11] Hagino, H., Oyama, M., & Sasaki, S. (2016). Laboratory testing of airborne brake wear particle emissions using a dynamometer system under urban city driving cycles. Atmospheric Environment, 131, 269-278.

[12] Amato, F., Schaap, M., Denvier van der Gon, H. A. C., Pandolfi, M., Alastuey, A., Kruken, M., et al. (2012). Effect of rain events on the mobility of road dust load in two Dutch and Spanish roads. Atmospheric Environment, 62, 352-358.

The Impacts of Urban Air Pollution on Malaysian Traffic Police: A Framework for Evaluation of Real-Time Monitoring System on Its Usability

P A S M Jamil[1], N A D M Yusof[1], K Karuppiah[1], I Rasdi[1], V How[1], S B M Tamrin[1], K K C Mani[2], H S Naeni[3], S Sambasivam[1], D H M S Nata[4]

[1]Department of Environmental and Occupational Health, Faculty of Medicine and Health Sciences, Universiti Putra Malaysia, 43400 UPM Serdang, Selangor, Malaysia
[2]Department of Community Health, Faculty of Medicine and Health Sciences, Universiti Putra Malaysia, 43400 UPM Serdang, Selangor, Malaysia
[3]Department of Industrial Engineering Design, School of Architecture and Environmental Design, Iran University of Science and Technology, Tehran, Iran
[4]Faculty of Industrial Science and Technology, Universiti Malaysia Pahang, 26300 Kuantan, Pahang, Malaysia

Abstract. The respiratory health of Malaysian Traffic Police has been compromised by working in heavy traffic and congested junctions with bad air for long hours. A wireless outdoor individual exposure device is vital to track their exposure, however, the efficacy of the system remains uncertain. While existing techniques exist to examine the efficacy of such system, there is a lack of methodology for engaging multiple assessment methods to evaluate the degree of user experience. This paper aims to propose a methodological framework tool for a quantitative evaluation of the wireless outdoor individual exposure indicator system prototype. A systematic search was conducted in major electronic databases (MEDLINE, Web of Science, Google), grey literature sources and all relevant data in the field. A three-stage framework consisting of simulation real-time monitoring, in-field testing, and usability testing is assembled. The three-stage framework proposed serves as a generic approach for evaluating the prototype with the purpose of tracking individual outdoor exposure. The method is capable of describing the complete evaluation process, from the accuracy and performance of the sensor to the extent of the end-user experience. Using the three-stage approach, future researchers may be able to create a monitoring system that is relevant to their needs.

1. Introduction

Rapid urbanisation in Malaysia has influenced the growth of our population, thus increasing the number of vehicles on roads [1]. As a consequence of increasing traffic on public roads, vehicular pollution has increased. Despite the emergence of better traffic management systems such as the Intelligent Transportation System (ITS) and Advanced Traffic Management Systems (ATMS), the need for travelling back and forth to work using personal vehicles is demanding [2]. To reduce traffic congestion, the Malaysian Traffic Police plays the main role by controlling and diverting the traffic at highly congested junctions and roads [3]. In highly dense areas, especially in the city centre of Kuala Lumpur, there is a higher need for traffic police to regulate the traffic flow [4]. They are more susceptible to adverse health effects such as respiratory diseases, heart diseases, and even lung cancer as they spend more time working at congested roads and junctions. [5] (Jamil et al., 2020a).

Unable to eradicate the cause of hazards to these traffic police, which is traffic-related air pollution, they would benefit from management control, such as an exposure indicator system [6]. Thus, a continuous real-time wireless outdoor individual exposure indicator system using low-cost sensors has a great potential to enhance the working environment in terms of safety and health for the Malaysian Traffic Police [7]. Numerous studies performed globally recommend the use of such a device to control air quality [8, 9, 10]. This device is practical and can effectively detect air pollutants that can help improve the protection of the population, personal health care, and environmental monitoring [11]. In recent times, these sensors have been used to collect high-resolution spatial and temporal air pollution data in real-time in many well-developed countries [12].

The development of low-cost air monitoring technologies has also been acknowledged in Europe and recommended to be included in the next Air Quality Directive [13]. Both the United States and the European Union (EU) have sponsored projects to assess low-cost air quality monitoring systems and set up test networks [14]. It shows that sensor technologies have a bright future and their use have proven to be beneficial. Nevertheless, air quality monitoring devices that are currently equipped with sensor technology are still limited in Malaysia. Similar devices with available sensors for wireless air monitoring have received a considerable amount of research in laboratory experiments and in-field testing globally [15,16]; however, their evaluation method on the extent of its usability is rarely documented. Assessments method for evaluating the effectiveness of wireless outdoor individual exposure indicator system prototype is still sparse in Malaysia.

Therefore, this study aims to test its accuracy, in-field performance, and, design a usability testing. Specifically, the study to propose a methodological framework tool for a quantitative evaluation of the wireless outdoor individual exposure indicator system prototype for traffic policemen.

2. Methods

A systematic search was conducted in major electronic databases (MEDLINE, Web of Science, Google). The characteristics of the studies that were used to construct methodological frameworks, were retrieved. The keyword used for the search was ‘outdoor air monitoring’, ‘in-field experimentation’, ‘wireless sensors’ and 'usability testing'. The search was conducted from 2000 until 2021.

The research methods and design, should be directed by the study question [17]. Thus, the evaluation of the system requires characterization against defined criteria that are relevant to the objective of the system and the extent of the reliability and performance of the system to be measured, as covered by the questions listed in the Table 1.

Table 1. Questions for the evaluation of a wireless outdoor system

Topic Area	Number	Question
Accuracy	1 (Q1)	Does the prototype provide sufficient accuracy of the readings when compared to a reference monitor?
In-field Performance	2 (Q2)	Does the prototype able to collect and transmit exposure data during in field experimentation (mobile testing)?
Usability	3 (Q3)	What is the usability score of the prototype for traffic policemen based on end-user feedback?

When developing a research plan, the research approach, depth of existing knowledge in the subject field, accessibility to data sources, and the availability of resources that would assist the implementation of the study should all be taken into account [17]. Based on research background, this section presents a three-stages framework to propose an evaluation method to determine the usability of the monitoring system. The summary of each stage according to the objectives is shown in Fig. 1.

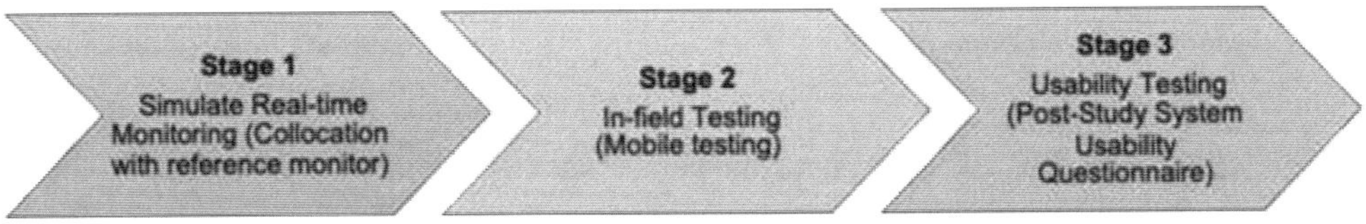

Figure 1. Proposed three-stage framework for evaluation of wireless outdoor individual exposure indicator system

3. Results and discussion

3.1. Stage 1: Simulate real-time monitoring (Collocation with reference monitor)

Q1 is concerned with the measurement of the sensor, its accuracy in the use of the wireless outdoor individual exposure indicator system. For an air quality monitoring system, it is essential to evaluate its effectiveness and sensitivity in collecting all the parameters desired to determine the surrounding air quality [18] (Wei et al., 2018). A simulation and comparison of real-time monitoring with a reference monitor side by side should prove the performance of the system [19]. The National Exposure Research Laboratory of United States Environmental Protection Agency has described a guideline entitled Instruction Guide and Macro Analysis Tool: Evaluating Low-Cost Air Sensors by Collocation with Federal Reference Monitors to effectively perform and evaluate any newly developed real time monitors [20]. In addition, the United Kingdom of Environmental Agency Technical Guidance Note is another available guideline widely used by developers around the globe for the evaluation method [21] (Environment Agency, 2011). By adopting the method in [20], it was proposed to compare the prototype and reference instruments by collocation and in this study, the temperature, relative humidity, $PM_{2.5}$, CO and NO_2 measurements of the prototype were compared with the measurements from DustTrak II for $PM_{2.5}$ and data from DOE for the rest of the parameters.

The collocation is a process by which a reference monitor and non-reference monitor (sensor) are operated at the same time and place under real world conditions for a defined evaluation period [20]. The regulated DOE air monitoring stations were indicated as reference instruments, with Federal Reference Method (FRM) or Federal Equivalent Method (FEM) methodologies being used. These instruments must meet the National Ambient Air Quality Standards (MAAQS). These instruments are referred to as "The Gold Standard" whereby most countries and company are using the FRM or FEM in their monitoring tools [22]. The DustTrak II Aerosol Monitor 8532, were used to measure $PM_{2.5}$ whereas the CO and NO_2 data from the prototype were compared using past year data from DOE. The data was collected and compared based on the concentration's uptake of the air pollutants. On this basis, the accuracy of the results from the prototype was observed through the variations when compared against the reference monitor [23] which is explained in detail in Table 2.

Table 2. Collocation with the reference monitor

Parameters	Reference Monitor	Performance Metrics
Particulate Matter 2.5 microns ($PM_{2.5}$)	DustTrak II Aerosol Monitor 8532	When evaluated at the same time and place, the variability in concentrations of all parameters from the system is compared to the reference monitor to assess if there is a significant difference between the two sensors. There should be less variability to reflect a good and accurate sensor.
Carbon Monoxide (CO)	DOE Air Monitoring Station	
Nitrogen Dioxide (NO_2)		

3.2. Stage 2: In-field testing (Performance)

Q2 is focused on the performance of the system in real practice to test the overall system. The system was proposed to be tested at the chosen study location. In this stage, the system was tested for its usage during the working hours of the respondents. The mobile testing was done in suburban district (Ampang Jaya) at Ampang Traffic Police Station. Both the prototype and the reference instruments were brought using a pickup truck and assembled at the tailgate. The mobile testing involves two hours driving around the same area with regular stopping at the traffic lights and junctions. The data was collected and observed for its performance in an ambient environment. The concentration measurement was plotted against time. Each parameter's average concentration was recorded and analysed descriptively. From this, the traffic police's working condition was portrayed and the exposure to outdoor air pollutants were known at the same time. The respondents were briefed on how to monitor the readings from the system using their smartphones. As they complete their working hours, they were asked to proceed to the next stage which is the usability testing (questionnaire).

3.3. Stage 3: Usability testing (Post-Study System Usability Questionnaire

The most important stage is to test its usability among the users. First, a usability testing plan is developed to evaluate the usability of the prototype (wireless outdoor individual exposure indicator system) for website and apps from the perspective of the Malaysian Traffic Police. The specific goals are to; identify the current level of exposure; search for previous exposure information by date and time; and save and download the information searched.

In usability testing, there is no specific formula for sample size agreed among the scholars for the best practice. Conversely, there is an established thumb rule which is a 4±1 or 10±2 model being used commonly as the usability testing sample size [24]. There are some valid reasons for using the thumb rule as the best approach for detecting problems in usability testing. Based on previous research [25,26,27,28], three major conclusions can be drawn: most problems are discovered by the first three to five participants; the increase in problem discovery after five participants is minimal; and the return on investment of usability testing can be maximised by reducing the sample size.

As stated before, for estimating the sample size required in the testing of a web interface, the rule of thumbs of 4±1 or 10±2 model is widely used to obtain a proportion of at least 80% of the problem discovery [24]. Experts in user experience studies agree that a smaller sample size is critical in understanding the core problem in a user experience study and achieving reasonable levels of problem recognition [29, 30,31,32, 33]. To avoid biased results from a small sample size, this study uses 10±2 as the rule of thumb. Hence, the total respondents for this study are 12 end users.

In the usability testing plan, a few test sessions are proposed in the following order, namely:

1. Screener test

This test was administered prior to the collection of data to classify respondents who meet the study criteria. The criteria for the users were working age at 18- 56 years old [34], have internet access and regular users of the internet [35]. The respondents were asked to complete the next stage afterwards.

2. Pre-test questionnaire (Parts A, B, C, D, and E)

In usability testing, the pre-test is essential to define the profile of user who will be performing the testing. A quick pre-test questionnaire is frequently undertaken before the user begins to experience the product. It is designed to assist the developer to better understand the user's views, goals, and behaviour in relation to the product under test [36,37,38]. In this test, the respondents were required to complete the questionnaire mainly on self-information, occupational information, knowledge of air quality, health effects from exposure to bad air, and actions taken to prevent exposure to bad air. These parts of the questionnaire are adopted from [39] which has been translated to Malaysian Language.

3. Post-task questionnaire (Task scenarios)

Next, the respondents were asked to install the application of this system on their smartphones and are required to follow the instructions given by the examiner according to the scenarios. This usability testing plan is adopted from [40].

4. Post-test questionnaire (Parts F, G, and H)

These parts of the questionnaire are adopted from the revised version three of standardized Post-Study System Usability Questionnaire (PSSUQ) by [41]. The questionnaire has also been used in other recent user-centered application research proving the usefulness in problem recognition of their product [42,43,44].

Overall, the questionnaire comprises a few parts, namely; Part A: Self-information, Part B: Occupational information, Part C: Knowledge of air quality, Part D: Health effects from exposure to bad air, Part E: Actions taken, Part F: User Feedback on the Prototype, Part G: Usability Testing (Apps), and Part H: Usability Testing (Website). The overall study flowchart from the early stage to the evaluation of the prototype is illustrated in Fig. 2.

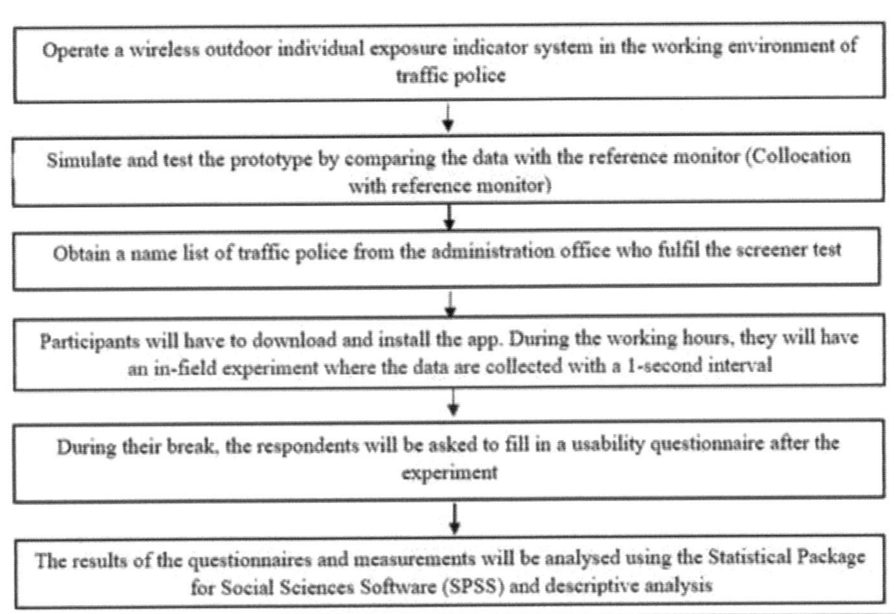

Figure 2. Flowchart of the proposed evaluation methods

4. Conclusion

This chapter has developed a generic methodology to evaluate the wireless outdoor individual exposure indicator system prototype with the aim of monitoring outdoor air quality. Furthermore, this method has laid the overall evaluation process from the sensor's accuracy and performance to the extent of end-user experience. Future researchers could seek to achieve a monitoring solution unique to their needs by following the three-stage framework.

5. References

[1] Department of Statistics Malaysia Compendium of Environment Statistics, Malaysia, 2017. Ministry of Economic Affairs, Department of Statistics, Malaysia. 2017.

[2] Yong M. Integrated Transport Information System (ITIS) for Kuala Lumpur: The clear way to go.

[3] Police Act 1967, Malaysia: Parliament

[4] Manogaran S, Ali M, Yusof KM, Suhaili R. Analysis of vehicular traffic flow in the major areas of Kuala Lumpur utilizing open-traffic. InAIP Conference Proceedings 2017 Sep 14 (Vol. 1883, No. 1, p. 020013). AIP Publishing LLC.

[5] Jamil PA, Karuppiah K, Rasdi I, How V, Tamrin SB, Mani KK, Sambasivam S, Yusof NA, Hashim NM. Associations of occupational, socio-demographic and lifestyle factors with lung functions in Malaysian traffic policemen. Annals of Global Health. 2020;86(1).

[6] Jamil PA, Yusof NA, Karuppiah K, Rasdi I, Jalalludin J, Tamrin SB, How V, Sambasivam S, Hashim NM. Malaysian Traffic Police in Highly Populated Areas: Is it Safe Working Outdoors on a Daily Basis?. Aerosol and Air Quality Research. 2020 Sep 1;20.

[7] Muhammad NS. Exposure to respirable dust (PM10) and respiratory health among traffic policemen

in Selangor. Advances in Environmental Biology. 2014; 8(15): 199–206.
[8] Kaivonen S, Ngai EC. Real-time air pollution monitoring with sensors on city bus. Digital Communications and Networks. 2020 Feb 1;6(1):23-30.
[9] Morawska L, Thai PK, Liu X, Asumadu-Sakyi A, Ayoko G, Bartonova A, Bedini A, Chai F, Christensen B, Dunbabin M, Gao J. Applications of low-cost sensing technologies for air quality monitoring and exposure assessment: How far have they gone?. Environment international. 2018 Jul 1;116:286-99.
[10] Castell, N., Dauge, F.R., Schneider, P., Vogt, M., Lerner, U., Fishbain, B., Broday, D.M. and Bartonova, A., 2017. Can commercial low-cost sensor platforms contribute to air quality monitoring and exposure estimates?.
[11] Tsow F, Forzani E, Rai A, Wang R, Tsui R, Mastroianni S, Knobbe C, Gandolfi AJ, Tao NJ. A wearable and wireless sensor system for real-time monitoring of toxic environmental volatile organic compounds. IEEE Sensors Journal. 2009 Oct 23;9(12):1734-40.
[12] Munir S, Mayfield M, Coca D, Jubb SA, Osammor O. Analysing the performance of low-cost air quality sensors, their drivers, relative benefits and calibration in cities—A case study in Sheffield. Environmental monitoring and assessment. 2019 Feb 1;191(2):94.
[13]Borrego C, Coutinho M, Costa AM, Ginja J, Ribeiro C, Monteiro A, Ribeiro I, Valente J, Amorim JH, Martins H, Lopes D. Challenges for a new air quality directive: the role of monitoring and modelling techniques. Urban Climate. 2015 Dec 1; 14:328-41.
[14] Woodall GM, Hoover MD, Williams R, Benedict K, Harper M, Soo JC, Jarabek AM, Stewart MJ, Brown JS, Hulla JE, Caudill M. Interpreting mobile and handheld air sensor readings in relation to air quality standards and health effect reference values: Tackling the challenges. Atmosphere. 2017 Oct;8(10):182.
[15] Pavani M, Rao PT. Urban air pollution monitoring using wireless sensor networks: A comprehensive review. International Journal of Communication Networks and Information Security. 2017 Dec 1;9(3):439-49.
[16] Jamil PA, Karuppiah K, Rasdi I, How V, Tamrin SB, Mani KK, Sambasivam S, Yusof NA. Mobile Outdoor Air Wireless Sensor Network for Exposure Monitoring among Traffic Policeman: A Review. Malaysian Journal of Medicine and Health Sciences. 2019; 15 Supp 2.
[17] Saunders M, Lewis P, Thornhill A. Research methods for business students. Pearson education; 2009.
[18] Wei P, Ning Z, Ye S, Sun L, Yang F, Wong KC, Westerdahl D, Louie PK. Impact analysis of temperature and humidity conditions on electrochemical sensor response in ambient air quality monitoring. Sensors. 2018 Feb;18(2):59.
[19] Li J, Zhang H, Chao CY, Chien CH, Wu CY, Luo CH, Chen LJ, Biswas P. Integrating low-cost air quality sensor networks with fixed and satellite monitoring systems to study ground-level PM2. 5. Atmospheric Environment. 2020 Feb 15; 223:117293.
[20] United States Environment Protection Agency. Instruction Guide and Macro Analysis Tool: Evaluating Low-Cost Air Sensors by Collocation with Federal Reference Monitors. 2019. Retrieved from https://www.epa.gov/airresearch/instruction-guide-andmacroanalysistoolevaluatinglow cost-air-sensorscollocation
[21] Environment Agency. Technical Guidance Note (Monitoring): Monitoring Ambient Air, Version 2. 2011. Retrieved from https://assets.publishing.service.gov.uk/government/uploads/system/uploads/attachment_data/file/301188/TGN_M8_Monitoring_Ambient_Air.pdf
[22] USEPA. Air Pollution Monitoring for Communities Grants. 2016.
[23] Kim S, Paulos E, Gross MD. WearAir: expressive t-shirts for air quality sensing. InProceedings of the fourth international conference on Tangible, embedded, and embodied interaction 2010 Jan 24 (pp. 295-296).
[24] Cazañas A, Parra E. Strategies for mobile web design. Enfoque UTE. 2017 Feb;8:344-57
[25] Virzi RA. Refining the test phase of usability evaluation: How many subjects is enough?. Human factors. 1992 Aug;34(4):457-68.

[26] Nielsen J, Landauer TK. A mathematical model of the finding of usability problems. InProceedings of the INTERACT'93 and CHI'93 conference on Human factors in computing systems 1993 May 1 (pp. 206-213).
[27] Lewis JR. Sample sizes for usability studies: Additional considerations. Human factors. 1994 Jun;36(2):368-78.
[28] Turner CW, Lewis JR, Nielsen J. Determining usability test sample size. International encyclopedia of ergonomics and human factors. 2006 Nov;3(2):3084-8.
[29] Majrashi K, Hamilton M, Uitdenbogerd AL, Al-Megren S. Cross-Platform Usability Model Evaluation. Multimodal Technologies and Interaction. 2020 Dec;4(4):80.
[30] Lestari S, Bahri S. Usability Web Analysis Using PSSUQ Method (Case Study on SME's Websites Application Fostered of Dispora Bandung Indonesia). International Journal of Psychosocial Rehabilitation. 2020 Apr 1;24(2).
[31] Zhou L, Bao J, Setiawan IM, Saptono A, Parmanto B. The mHealth APP usability questionnaire (MAUQ): development and validation study. JMIR mHealth and uHealth. 2019 Apr 11;7(4):e11500.
[32] Cho H, Powell D, Pichon A, Thai J, Bruce J, Kuhns LM, Garofalo R, Schnall R. A mobile health intervention for HIV prevention among racially and ethnically diverse young men: usability evaluation. JMIR mHealth and uHealth. 2018;6(9):e11450.
[33] Bernacki J, Błażejczyk I, Indyka-Piasecka A, Kopel M, Kukla E, Trawiński B. Responsive web design: testing usability of mobile web applications. InAsian Conference on Intelligent Information and Database Systems 2016 Mar 14 (pp. 257-269). Springer, Berlin, Heidelberg.
[34] Wan Ahmad WI, Ismail Z. Some demographic aspects of population aging in Malaysia. World Applied Sciences Journal. 2014;30(7):891-4.
[35] Carvalho MC, Dias FS, Reis AG, Freire AP. Accessibility and usability problems encountered on websites and applications in mobile devices by blind and normal-vision users. InProceedings of the 33rd Annual ACM symposium on applied computing 2018 Apr 9 (pp. 2022-2029).
[36] Boothe C, Strawderman L, Hosea E. The effects of prototype medium on usability testing. Applied ergonomics. 2013 Nov 1;44(6):1033-8.
[37] Broekhuis M, van Velsen L, Peute L, Halim M, Hermens H. Conceptualizing Usability for the eHealth Context: Content Analysis of Usability Problems of eHealth Applications. JMIR Formative Research. 2021 Jul 27;5(7):e18198.
[38] Hernandez H. The art of asking questions in usability testing. V Akendi.
[39] Karki K, Sushmita KC, Neupane S. Prevention of respiratory problems among traffic police: A cross sectional study in Kathmandu valley exploring knowledge and practice. Lung cancer. 2018; 101:60-8.
[40] Barnum CM. Usability testing essentials: ready, set... test!. Morgan Kaufmann; 2020 Jun 27.
[41] Lewis JR. Psychometric evaluation of the PSSUQ using data from five years of usability studies. International Journal of Human-Computer Interaction. 2002 Sep 1;14(3-4):463-88.
[42] Farao J, Malila B, Conrad N, Mutsvangwa T, Rangaka MX, Douglas TS. A user-centred design framework for mHealth. PloS one. 2020 Aug 19;15(8):e0237910.
[43] Callan JA, Jacob JD, Siegle GJ, Dey A, Thase ME, Dabbs AD, Kazantzis N, Rotondi A, Tamres L, Van Slyke A, Sereika S. CBT MobileWork©: User-centered development and testing of a mobile mental health application for depression. Cognitive Therapy and Research. 2021 Apr;45(2):287-302.
[44] Ng MM, Firth J, Minen M, Torous J. User engagement in mental health apps: a review of measurement, reporting, and validity. Psychiatric Services. 2019 Jul 1;70(7):538-44.

Acknowledgments
Acknowledgement is given to the Royal Malaysian Police, specifically Officer in Charge of Kuala Lumpur Traffic Police Station and the respondents of this study.

Indoor air quality (IAQ) in a naval ship after refit program: a time variation analysis

Maryam Zahaba[1], Nur Sarah Fatihah Tamsi[1], Sakinah Engliman[2], Amirul Faiz Kamaruddin[2, 3], Noor Artika Hassan[4] and Arman Ariffin[5]

[1]Chemistry Department, Kulliyyah of Science, International Islamic University Malaysia, 25200 Kuantan, Malaysia
[2]Biotechnology Engineering Department, Kulliyyah of Engineering, International Islamic University Malaysia, 53100 Gombak, Malaysia
[3]Faculty of Engineering, KD PELANDOK, Royal Malaysian Navy Naval Base, 32100 Lumut, Malaysia
[4]Community Medicine Department, Kulliyyah of Medicine, International Islamic University Malaysia, 25200 Kuantan, Malaysia
[5]Eastern Maintenance Overseer Unit - Royal Malaysian Navy, RMN Naval Base, Tg Purun, 87000 W.P. Labuan, Malaysia

Abstract. Refurbishments of the ship's external and internal structures are the main scopes of a refit program. These activities may affect the indoor air quality (IAQ) inside ships and increase the indoor air pollutants (IAP) concentrations onboard. Therefore, continuous IAQ monitoring is needed to determine IAP exposure to the ship's crew. This study evaluates the changes in IAQ conditions inside a naval ship over a two-time interval to determine the effect of compliance with the recommended engineering control measures proposed in the first assessment. Following the standard of the Industry Code of Practice on Indoor Air Quality 2010 (ICOP on IAQ 2010), seven IAQ parameters (temperature, relative humidity (RH), carbon dioxide (CO_2), respirable particulates/particulate matter (PM_{10}), total volatile organic compounds (TVOC), bacterial count, and fungal count) were measured in two assessment phases. The first phase was conducted after the ship completed the refit program, and the second phase began three months later, following the execution of the recommended engineering control measures. According to the findings of this study, all IAQ parameters improved when compared to the first phase assessment. However, some of the readings were still non-compliance with the standards of ICOP on IAQ 2010. In conclusion, the ship's IAQ parameters were improved following the recommended engineering control measures, although more enhanced approaches were required to ensure all parameters complied with the ICOP on IAQ 2010.

Keywords: Indoor air quality (IAQ); Naval ship; Engineering control; Ship refit program.

[1] Corresponding author: maryamzahaba@iium.edu.my

1. Introduction

Air pollutants may cause adverse effects on the environment and health. The adverse effects of indoor air pollutants (IAP) can be more critical than those effects caused by the outside air pollutants [1]. Previous studies have proved that human exposure to the IAP might be two to multiple times, in terms of the numerous harmful IAP concentrations, higher than the outside air pollutants [2]. On the other hand, low IAP concentrations indicate good indoor air quality (IAQ), leading to a positive impact on the health and productivity of the occupants [3].

Moreover, IAQ is an essential component of indoor environmental quality (IEQ) and plays an important role in ensuring a healthy and safe workplace for the workers and occupants, including those crews in ships. Poor IAQ in onboard ships may cause discomfort and health risks to the crews besides causing equipment breakdown [4]. Ships undergo several refitting programs, with major external and internal refurbishment throughout their service life [5]. Activities such as repainting, adding new furnishings, major component overhauls, and redesigning the interior spaces are among the work scopes of the refitting program [6]. These activities may affect the ship's IAQ and increase the concentrations of IAP onboard. Previously published studies are limited to local surveys with localized climate environments and specific ships condition. A study by Österman et al. (2016) explained the effects of IAQ chemical parameters on passenger ferries in temperate climate countries that specifically operate in the same environment and will not adequately represent other vessels [7]. Meanwhile, few authors have been able to draw on any systematic research into this area of studies that conduct specific IAQ research and compare it to local IAQ guidelines [8][9]. Nonetheless, little is known about the IAQ onboard ships in various conditions, and there is an urgent need to address the safety and health-related problems caused by poor IAQ onboard ships. Thus, this study was done to evaluate the changes in IAQ conditions inside a naval ship over a two-time interval to determine the effect of adhering to the recommended engineering control measures proposed in the first assessment.

2. Methodology

2.1 Sampling sites and assessment phases

The first assessment phase was conducted at the end of December 2020, and the second at the beginning of March 2021. During the first assessment, the ship had recently undergone a refitting program, which included repainting, redesigning the interior spaces, installing new furnishings, and changing other major and minor components. At the end of the assessment, several recommendations and engineering control measures were recommended to the crews to improve the IAQ inside the ship. The second assessment phase was conducted in continuity after three months with the execution of the suggestions made in the previous assessment. The sampling points involved the ship's ten compartments, which comprise accommodation and working spaces. Both assessment phases were conducted when the ship was alongside the jetty. The summary of the IAQ assessment's flow is shown in Figure 1:

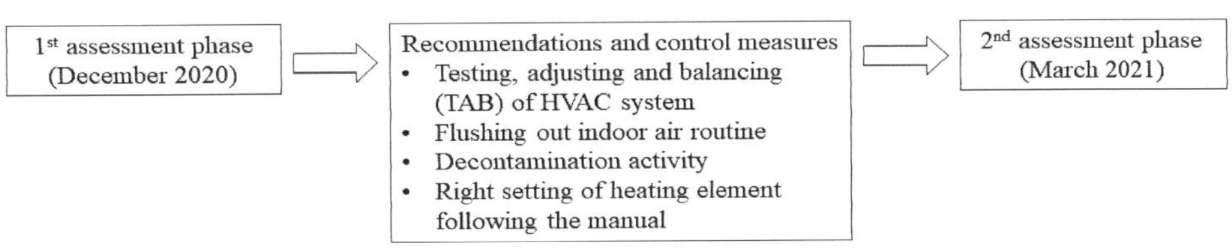

Figure 1. The flow of IAQ assessments within two-time intervals and recommendations suggested from the first sampling phase to improve IAQ onboard

2.2 Sampling campaign and instrumentation

The IAQ parameters inside the ship compartments were measured following the Industry Code of Practice on Indoor Air Quality 2010 (ICOP on IAQ 2010) by the Department of Occupational Safety and Health [10]. Prior to the sampling activity, a walkthrough inspection was conducted, including the semi-formal meeting, walkthrough observation and occupant's interview and questionnaire administration. This study measured physical and chemical parameters such as temperature (°C), relative humidity (RH), and carbon dioxide (CO_2) which were measured using the TSI VelociCalc 9565. Total volatile organic compounds (TVOC) concentrations were measured using the RAE Systems ppbRAE, and respirable particulates/particulate matter (PM_{10}) concentrations were measured using the Kanomax Digital Dust Monitor Model 3443. The eight-hour sampling period was divided into four slots, with a direct reading taken for five minutes intervals within 30 minutes period of sampling duration for each slot.

Besides that, the bacterial and fungal counts were also assessed as biological parameters for the IAQ. The airborne bacteria and fungi were sampled using the PBI International SAS Super IAQ on trypticase soy agar (TSA) for bacteria and malt extract agar (MEA) for fungi. The sampling was conducted in triplicate for a minute at a flow rate of 100 L/m following the National Institute of Safety and Health (NIOSH) Manual Analytical Method (NMAM) 0800 [11].

2.3 Data analysis

The average total bacterial and fungal count was determined by calculating the colony-forming unit per cubic meter (CFU/m³) from the previous sampling culture media using the formula (1).

$$\text{Total bacterial or fungal count, CFU/m}^3 = \frac{Total\ count\ X\ 1000}{Volume\ of\ sampled\ air} \quad (1)$$

The 8-hour total weighted average (TWA_8) of the IAQ parameters was calculated to estimate the average readings of the 8-hour exposure in the compartments. The TWA_8 readings were then compared to the acceptable limit from the ICOP on IAQ 2010. Based on the normality test on the sample data, a non-parametric Wilcoxon signed-rank test was used to analyze the comparison of TWA_8 readings for physical and chemical parameters of each ship's compartment between the two phases of assessment.

3. Results and discussions

3.1 Physical parameters

Data in Table 1 shows that the temperature readings during the second assessment phase were significantly reduced in all compartments except electrical room 2 ($Z = -0.715$, $P = 0.475$). However, the temperature drop from the first phase is too low and falls below the standard set by the ICOP on IAQ 2010 (23–26°C). This result can be explained by the fully functional heating, ventilation, and air conditioning (HVAC) system and execution of the testing, adjusting, and balancing (TAB) done by the engineering department as recommended in the first phase assessment. The RH readings in all sampling points show a positive change from the previous assessment as the temperatures drop. As shown in Table 1, all RH levels in the compartments were within the standard's acceptable limit. These factors could explain the relatively good correlation between temperature and RH [12]. In addition, the heating element was set according to the manufacturer's recommendations in order to control the humidity onboard, and the heating setting was adjusted after the first assessment.

Table 1. TWA_8 reading for physical parameters.

Sampling point	Temperature, °C				Relative humidity, %			
	1st phase	2nd phase	Z score	p value	1st phase	2nd phase	Z score	*P*-value
Galley	29.1 ± 0.2[a]	24.8 ± 1.1	-4.293[b]	< 0.001[c]	72.3 ± 4.0[a]	55.0 ± 2.7	-4.286[b]	< 0.001[c]
Common room 1	26.7 ± 0.3[a]	19.3 ± 0.5[a]	-4.291[b]	< 0.001[c]	75.2 ± 2.5[a]	68.2 ± 2.1	-4.287[b]	< 0.001[c]
Officer's cabin 1	25.6 ± 0.6	20.2 ± 0.8[a]	-4.291[b]	< 0.001[c]	73.5 ± 3.8[a]	61.3 ± 1.5	-4.287[b]	< 0.001[c]
Common room 2	26.6 ± 0.3[a]	20.0 ± 0.4[a]	-4.294[b]	< 0.001[c]	75.2 ± 2.2[a]	65.4 ± 1.2	-4.286[b]	< 0.001[c]
Officer's cabin 2	25.6 ± 0.6	22.6 ± 0.1[a]	-4.291[b]	< 0.001[c]	66.0 ± 4.1	68.3 ± 3.3	-2.029	0.042[c]
Gym	29.3 ± 0.9[a]	27.1 ± 1.4[a]	-4.305[b]	< 0.001[c]	72.0 ± 4.2[a]	57.1 ± 5.2	-4.286[b]	< 0.001[c]
Electrical room 1	25.0 ± 1.5	22.7 ± 0.6[a]	-3.987[b]	< 0.001[c]	67.1 ± 7.0	64.0 ± 3.3	-1.958[b]	0.05
Electrical room 2	19.5 ± 2.3[a]	21.3 ± 10.4[a]	-0.715	0.475	69.1 ± 3.4	64.2 ± 4.5	-2.972[b]	0.003[c]
Officer's cabin 3	23.9 ± 0.9	21.0 ± 1.0[a]	-4.299[b]	< 0.001[c]	67.9 ± 2.8	66.4 ± 3.0	-2.788[b]	0.005[c]
Bridge	25.7 ± 0.6	22.1 ± 1.5[a]	-4.289[b]	< 0.001[c]	70.2 ± 13.3[a]	64.4 ± 3.1	-2.029[b]	0.043[c]

[a] Noncompliance reading with ICOP on IAQ 2010
[b] 1st phase assessment > 2nd phase assessment
[c] Significant different at $P < 0.05$

Table 2. TWA_8 reading for chemical parameters

Sampling point	Carbon dioxide (CO_2), ppm				Total volatile organic compound (TVOC), ppm				Particulate matter (PM_{10}), mg/m^3			
	1st phase	2nd phase	Z score	*P*-value	1st phase	2nd phase	Z score	*P*-value	1st phase	2nd phase	Z score	*P*-value
Galley	890 ± 99	754 ± 36	-3.743[b]	< 0.001[c]	19.30 ± 10.33[a]	4.09 ± 0.81[a]	-4.286[b]	< 0.001[c]	0.108 ± 0.032	0.007 ± 0.010	-4.258[b]	< 0.001[c]
Common room 1	1092 ± 117[a]	761 ± 97	-4.286[b]	< 0.001[c]	36.27 ± 24.45[a]	4.13 ± 0.29[a]	-4.286[b]	< 0.001[c]	0.072 ± 0.024	0.002 ± 0.002	-4.287[b]	< 0.001[c]
Officer's cabin 1	1229 ± 90[a]	913 ± 145	-4.143[b]	< 0.001[c]	12.76 ± 2.9[a]	2.96 ± 0.18	-4.286[b]	< 0.001[c]	0.040 ± 0.018	0.004 ± 0.002	-4.287[b]	< 0.001[c]
Common room 2	1126 ± 83[a]	816 ± 81	-4.286[b]	< 0.001[c]	16.37 ± 3.96[a]	4.27 ± 0.26[a]	-4.286[b]	< 0.001[c]	0.154 ± 0.230[a]	0.003 ± 0.007	-4.287[b]	< 0.001[c]
Officer's cabin 2	1247 ± 328[a]	713 ± 93	-4.287[b]	< 0.001[c]	8.70 ± 2.13[a]	3.27 ± 0.14[a]	-4.286[b]	< 0.001[c]	0.314 ± 0.306[a]	0.002 ± 0.001	-4.288[b]	< 0.001[c]
Gym	911 ± 379	781 ± 58	-1.571[b]	0.116	5.03 ± 4.13[a]	3.50 ± 0.37[a]	-1.209[b]	0.304	0.235 ± 0.200[a]	0.004 ± 0.001	-4.287[b]	< 0.001[c]
Electrical room 1	2040 ± 404[a]	977 ± 172	-4.286[b]	< 0.001[c]	14.19 ± 1.92[a]	4.02 ± 0.39[a]	-4.286[b]	< 0.001[c]	0.035 ± 0.037	0.005 ± 0.005	-4.289[b]	< 0.001[c]
Electrical room 2	1008 ± 294[a]	960 ± 223	-1.243[b]	0.214	4.93 ± 2.02[a]	3.20 ± 0.16[a]	-2.4[b]	0.016[c]	0.022 ± 0.045	0.045 ± 0.003	-2.645	0.008[c]
Officer's cabin 3	1213 ± 203[a]	987 ± 139	-2.286[b]	0.022[c]	5.14 ± 1.42[a]	3.42 ± 0.26[a]	-4.256[b]	< 0.001[c]	0.022 ± 0.006	0.016 ± 0.008	-3.304[b]	0.001[c]
Bridge	1265 ± 169[a]	828 ± 156	-4.172[b]	< 0.001[c]	1.47 ± 0.56[a]	1.79 ± 0.34	-1.743	0.081	0.063 ± 0.146	0.004 ± 0.002	-4.287[b]	< 0.001[c]

[a] Noncompliance readings with ICOP on IAQ 2010
[b] 1st phase assessment > 2nd phase assessment
[c] Significant different at $P < 0.05$

3.2 Chemical parameters

In the case of chemical parameters, a Wilcoxon signed-rank test in Table 2 shows that the difference in CO_2 concentration between the first and second assessments was statistically significant in all compartments except the gym (Z = -1.571, P = 0.116) and electrical room 2 (Z = -1.243, P = 0.214). All compartments improved their ventilation by demonstrating a decreasing CO_2 concentration in the second assessment phase and falling below the acceptable limit set by the ICOP on IAQ standard (< 1000 ppm). The data from the gym and electrical room 2 must be interpreted with caution because the source of the CO_2 during sampling could be human respiration [13]. Overall, the decreasing CO_2 trend could be attributed to the engineering control measures activities conducted by ship crews. To reduce the CO_2 concentration, fresh air must be introduced into the compartment, causing the concentrated IAP to dilute away. According to Fang et al. (2019), forced ventilation is typically used in tunnel construction because it is a cost-effective method to provide large amounts of fresh air. A similar outcome in this study was found to be significant when compared to previous findings [14].

Furthermore, Table 2 shows a statistical difference in TVOC concentration between the first and second assessment in all compartments except the gym (Z = -1.209, P = 0.304) and the bridge (Z = -1.743, P = 0.081). Apart from the bridge, all readings showed a decrease in TVOC concentration during the second assessment. All compartments recorded exceeded the value of TVOC during the first assessment due to the major and minor refurbishment activity of the ship, as expected from the refitting program. This is supported by the idea that new furniture, painting, and surrounding materials such as carpets cause TVOC emissions [15]. Despite some inconsistency, the concentrations of TVOC were interestingly reduced from the initial IAQ assessment. Presumably, it was due to routinely flush out activity taken by the crew as recommended in the first assessment. The three months of an activity practice, however, was insufficient to dilute the TVOC emission, as most of the compartments still recorded an exceeding concentration from the acceptable limit as prescribed in the ICOP on IAQ 2010.

The statistical analysis results for PM_{10}, similar to other parameters, show a significant difference in concentration between the first and second phase assessments in all compartments. All readings showed positive feedback from the engineering control measures taken, which routinely flushes out the indoor air. The ship's refurbishment activity indeed increased the concentration of PM_{10}, as seen in the common room 2, officer's cabin 2, and gym during the first assessment. According to studies, renovation and refurbishment activities increase the concentration of PM_{10} in the surrounding air [16]. Since PM_{10} can adversely affect the health of the ship's crew, it can be avoided further by employing the individual prevention technique of wearing personal protective equipment (PPE) during the activity. A good practice of wearing appropriate PPE while working in areas with high concentrations of PM_{10} can reduce the risk of human inhalation of PM_{10}. This corroborates the ideas of Shakya et al. (2017) who investigated the efficacy of facemasks in reducing particulate matter exposure [17].

3.3 Biological parameters

Most compartments reported a decrease in airborne bacterial and fungal count from the first phase assessment to the second assessment, except for fungal count in the bridge. Figures 2 and 3 show the different results for the bioaerosol count between the two phases and demonstrate the efficiency of the decontamination activity and the right setting of humidity controller on the HVAC system as practiced by the crew, by referring to the manufacturer's manual. The refurbishment activity during the refitting program had caused the resuspension of the microorganisms in the air and explained the exceeding CFU for the bacteria and fungi during the first assessment. Mousavi et al. (2020) found that renovation and refurbishment activities increase the resuspension of PM as well as microorganisms since PM can be a potential carrier for the microorganisms [18]. High CFU count of microbial is related to the high concentration of PM [19][20].

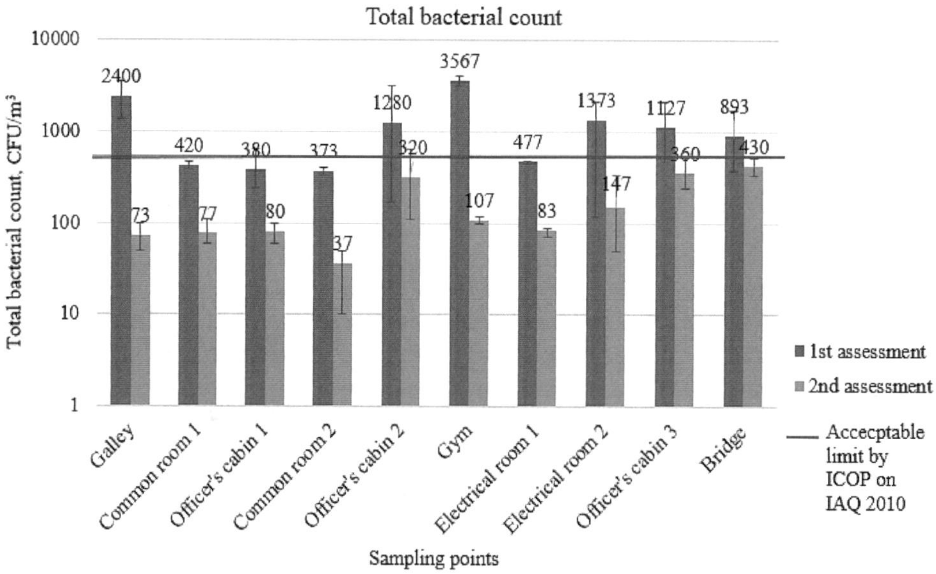

Figure 2. Total bacterial count during assessment phases

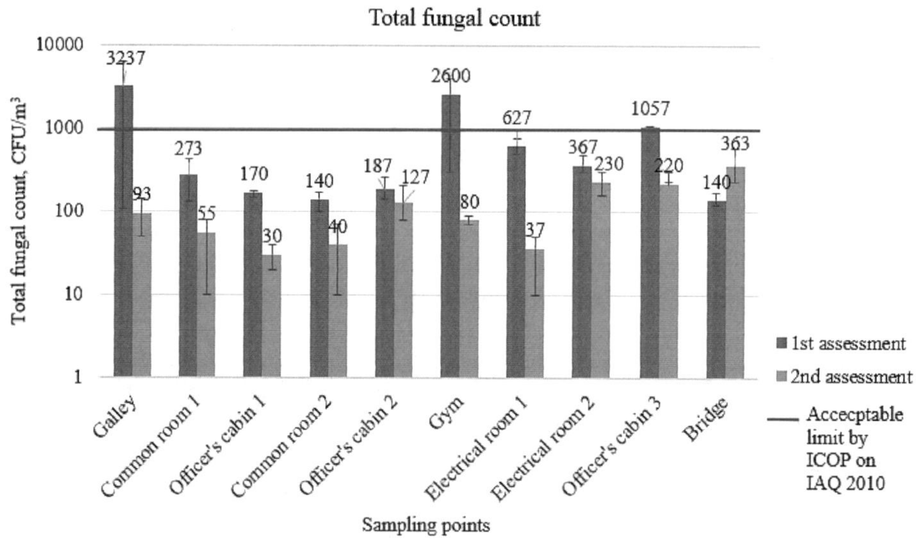

Figure 3. Total fungal count during assessment phases

4. Conclusions

Overall, this study found that the IAQ parameters improved after a three-month interval of assessment and execution of recommended control measures from the first assessment. The findings demonstrate the effectiveness of activities such as TAB of HVAC system, routinely flushing out the indoor air and decontamination activities, as well as the proper setting of heating elements to control humidity, in reducing IAP concentrations and the dilution process over the months. Even though this study was based on limited samples on a ship, the findings proved that proper maintenance and a high level of awareness of IAQ among the crews onboard would sustain a healthier environment inside the ship. Hence, further studies regarding the IAQ onboard ships would be worthwhile.

5. Acknowledgement
This study would like to thank the International Islamic University of Malaysia (IIUM) and the Royal Malaysian Navy (RMN) for their assistance with this work (grant ID: RMCG20-044-0044). We would like to express our sincere gratitude to the Eastern Maintenance Overseer Unit, RMN, for granting us access to the ship, as well as all of the crews who were involved in the study.

6. References

[1] Tsakas M P, Siskos A P, Siskos P A 2011 Indoor air pollutants and the impact on human health. *Chemistry, Emission Control, Radioactive Pollution and Indoor Air Quality*, 447-84.

[2] Marć M, Śmiełowska M, Namieśnik J, Zabiegała B 2018 Indoor air quality of everyday use spaces dedicated to specific purposes—a review. *Environmental Science and Pollution Research,* 25(3), 2065-82.

[3] Loftness V 2020. Sustainable built environments. Haase D, editor. New York: Springer; 2020.

[4] Rozzi-ochs J A 2006 *Shipboard Ventilation Systems and Design Standards on Board*

[5] Liu Y, Frangopol D M and Cheng M 2019 Risk-informed structural repair decision making for service life extension of aging naval ships. *Marine Structures*, *64*, 305–321. https://doi.org/10.1016/j.marstruc.2018.10.008

[6] Cruise 2020 *Top 10 cruise ship refit and refurbishments in 2020* https://cruise-arabia.com/2020/02/14/top-10-cruise-ship-refit-and-refurbishments-in-2020/

[7] Österman C, Langer S and Moldanova J 2016 Personal exposure of seafarers to air pollutants and perceived indoor air quality on a passenger ferry. *Indoor Air 2016. The 14th International Conference on Indoor Air Quality and Climate.*

[8] Langer S, Moldanová J and Österman C 2016 Influence of fuel change on indoor environmental quality on-board a passenger ferry. *The 14th International Conference of Indoor Air Quality and Climate, Ghent, Belgium July 4-8. International Society of Indoor Air Quality and Climate-ISIAQ*

[9] Carrer P and Wolkoff P 2018 Assessment of indoor air quality problems in office-like environments: Role of occupational health services. *International Journal of Environmental Research and Public Health*, *15*(4). https://doi.org/10.3390/ijerph15040741

[10] Department of Safety and Health 2010 *Industry Code of Practice on Indoor Air Quality 2010.* Ministry of Human Resources Malaysia

[11] Lonon M K 1998 *Bioaerosol sampling (Indoor Air) 0800 Culturable organisms : bacteria , fungi , thermophilic actinomycetes* (Issue 1).

[12] Pei C. Ou, Q and Pui, D Y H 2021 Effects of temperature and relative humidity on laboratory air filter loading test by hygroscopic salts. *Separation and Purification Technology*, 117679. https://doi.org/10.1016/j.seppur.2020.117679

[13] Andrade A, Dominski F H, Pereira M L, de Liz C M and Buonanno G 2018 Fitness centers demonstrate CO_2 concentration levels above recommended standards. *Acta Scientiarum - Health Sciences*, *40*. https://doi.org/10.4025/actascihealthsci.v40i1.35768

[14] Fang Y, Yao Z and Lei S 2019 Air flow and gas dispersion in the forced ventilation of a road tunnel during construction. *Underground Space (China)*, *4*(2), 168–179. https://doi.org/10.1016/j.undsp.2018.07.002

[15] Mečiarová L, Vilčeková S, Burdová E K and Kiselák J 2017 Factors effecting the total volatile organic compound (TVOC) concentrations in slovak households. *International Journal of Environmental Research and Public Health*, *14*(12). https://doi.org/10.3390/ijerph14121443

[16] Szewczyńska M, Dobrzyńska E, Pośniak M. Determination of phthalates in particulate matter and gaseous phase emitted in indoor air of offices. *Environmental Science and Pollution Research.* 2020 Sep 22:1-9.

[17] Shakya K M, Noyes A, Kallin R and Peltier R E 2017 Evaluating the efficacy of cloth facemasks in reducing particulate matter exposure. *Journal of Exposure Science and Environmental Epidemiology*, *27*(3), 352–357. https://doi.org/10.1038/jes.2016.42

[18] Mousavi E, Sharma V, Gajjar D and Naini S S 2020 Renovation in hospitals: a case study on the use of control cubes for local repairs in health-care facilities. *Journal of Facilities Management.*
[19] Khamal R, Isa Z M, Sutan R, Noraini N M R and Ghaz H F 2019 Indoor particulate matters, microbial count assessments, and wheezing symptoms among toddlers in urban day care centers in the district of seremban, Malaysia. *Annals of Global Health*, *85*(1), 1–12. https://doi.org/10.5334/aogh.2425
[20] Morakinyo O M, Mokgobu M I, Mukhola M S and Godobedzha T 2019 Biological composition of respirable particulate matter in an industrial vicinity in South Africa. *International Journal of Environmental Research and Public Health*, *16*(4). https://doi.org/10.3390/ijerph16040629

An overview of indoor air pollution in the Malaysian kindergarten environment

I B Zakaria[1,2] and N Mahyuddin[1*]

[1] Department of Building Surveying, Faculty of Built Environment, University Malaya, 50603 Kuala Lumpur, Malaysia and Centre for Building, Construction and Tropical Architecture (BuCTA), Faculty of Built Environment, University Malaya, 50603 Kuala Lumpur, Malaysia
[2] Programme of Building Surveying, Department of Built Environment Studies & Technology, Faculty of Architecture, Planning and Surveying, Universiti Teknologi MARA, Perak Branch, Seri Iskandar Campus, 32610 Seri Iskandar, Perak, Malaysia
* Corresponding author: hayati@um.edu.my

Abstract. The components of indoor air determine the quality of the indoor environment, which affects the health and well-being of inhabitants. Exposure to high levels of indoor air pollution in a kindergarten environment has a detrimental influence on children's learning performance and increases the risk of respiratory problems that could lead to absenteeism. This paper explores and summarises the literature on indoor air pollution in kindergartens in the Malaysian environment. A review was done by analyzing papers taken from four electronic databases: Scopus, Science Direct, PubMed and Google Scholar. Information on indoor pollution levels and determinants sources were extracted from 17 studies. The most investigated pollutants were carbon dioxide (CO_2), particulate matter ($PM_{2.5}$ and PM_{10}), volatile organic compounds (VOCs), carbon monoxide (CO), fungi and bacteria. Inadequate ventilation systems and overcrowded classrooms all contributed to excessive CO_2 levels. Indoor $PM_{2.5}$ sources are generated from cooking activities while cleaning, opening windows and movement activities of children were the primary sources of coarse particles. High concentrations of VOCs are emitted from a wide variety of indoor sources. Mobile vehicles and the proximity of kindergartens to busy roads were key contributors to higher CO concentrations. The findings highlight the importance of interventions to improve indoor air pollution in kindergarten premises in various settings.

1. Introduction

In modern society, people are exposed to various pollutants every day regardless of their locations and activities. The exposure depends on several factors, such as the surrounding environment and the individual's lifestyle. In 2017, 55% of the world's population lived in urban settings, representing 4.1 billion people, compared to only 25% in 1960 [1]. Due to increased urbanization and industrialization, air pollution has become a major environmental issue in low and middle-income countries. In 2016, almost half of the world's population was exposed to air pollution levels of at least 2.5 fold above WHO's safety standard [2].

Changes in outdoor concentrations of air pollutants have been well documented over the last few decades; however, the effects of air pollution on an individual's health are proportional to their total personal exposure (outdoor and indoor) [3]. Growing scientific evidence indicates that the air inside a building can be more contaminated than the air outside [4]. Exposure to indoor air pollution will impact health, especially the elderly, adolescents and children who spend most of their time indoors [5]. Exact pollution concentration can result in a higher burden on children's bodies than adults because they inhale a larger amount of air relative to their body weights [6].

Besides home, a kindergarten is a place where children spend more time indoors on a weekly basis. Therefore, the learning environment has to be more conducive for children to learn and develop. Previously, kindergarten quality generally concerned itself with social and curricular aspects of early childhood education programs. Less attention has been paid to the physical environment of kindergartens [7], where it is known to influence children's learning performance, health and behavior [7]–[9]. According to research conducted in the United States [10], most early childhood education programs do not provide appropriate indoor environments, which will lead to the increment of indoor air pollution. From a broader perspective, there is a need to collaborate research on indoor air pollution with physical setting requirements [11] without compromising early childhood education requirements.

Knowledge of indoor air pollution in Malaysian kindergarten is still at the infancy stage despite the increasing demand for kindergarten every year [12]. Lack of reviews on indoor air pollution in kindergarten environments has caused the absence of comprehension of existing literature. In order to develop a conducive environment with less indoor air pollution, indoor air problems in kindergarten must be studied. Therefore, this study aims to fill the knowledge gap by systematically reviewing related articles published between 2011 until 2020 and describing the findings for better understanding. The following research question guides this review: What are the main indoor air problems in Malaysian kindergartens?

2. Indoor Air Pollution

Indoor air quality is considered acceptable is when there is no contaminant at a harmful concentration level, and when exposed to it, most occupants (80% or more) do not express dissatisfaction[13]. In a building, the quality of indoor air is determined by the amount of air pollutants present, classified into two groups: chemical and biological. Under chemical subgroup: solid particles, liquid droplets and gas [14]. The amount of air pollutants in indoor space does not remain constant. It fluctuates according to changes in building operation, human activity and the outdoor environment.

Over the last decade, numerous studies have been undertaken to assess indoor air quality, mainly in residential, commercial and school environments [15]–[19]. A substantial number of indoor air pollutions were identified and quantified, including volatile organic compounds (VOCs), particulate matter (PM), sulfur dioxide (SO_2), nitrogen oxides (NOx), ozone (O_3), carbon monoxide (CO), carbon dioxide (CO_2), bioaerosols and selected heavy metals. Few authors in developed countries reported indoor air quality in kindergartens [20]–[24], and even fewer in developing countries. Previous research has shown that outdoor air has a stronger impact on indoor air quality in tropical and subtropical climates than in cold or temperate climates, resulting in differing impacts on the indoor environment [25].

2.1 Malaysian Air Quality Standard

In Malaysia, the Department of Safety and Health of the Ministry of Human Resources produced an Industry Code of Practice on Indoor Air Quality (ICOP) launched in August 2010 [26] and replaced the previous 2005 Code of Practice. This code of practice applies to any enclosed space served by a mechanical ventilating and air conditioning (MVAC) system, including those with air-cooled split units. These general duties are governed under the Occupational Safety and Health Act 1994 (Act 514). Although this code of practice is not intended for use in a kindergarten context, it can be used as a guide for determining the concentration limits permissible in an indoor environment. A list of indoor contaminants and the permissible limits are shown in Table 1.

Table 1. List of indoor air contaminants and the maximum limits

Indoor Air Contaminant	Eight-hour time-weighted average airborne concentration		cfu/m3
	ppm	mg/m3	
Carbon dioxide	C1000		
Carbon monoxide	10		
Formaldehyde	0.1		
Respirable particulates		0.15	
Total volatile organic compounds (TVOC)	3		
Total bacteria counts			500
Total fungal counts			1000

3. Materials and method

This systematic literature review (SLR) employ numerous unique procedures which are different from traditional review. It is guided using the publication standard of PRISMA (Preferred Reporting Items for Systematic Reviews and Meta-Analyzes)[27]. The review was systematic because it seeks to answer formulated questions and uses systematic methods to identify, select, and critically evaluate research. It can be replicated by other researchers that use a similar approach [28]. PRISMA assists authors in precisely tracking literature work according to the research question via the four procedures of identification, screening, eligibility and inclusion [29].

3.1 Literature searches

For this study, the following key data were extracted from the electronic databases of Elsevier Bibliographic Database (SCOPUS), Science Direct, PubMed (MEDLINE) and Google Scholar. Articles published between 2011 and 2020 were covered in our search. We concentrated on measured exposures to chemical air pollutants, particles and bioaerosol pollutants in the kindergarten indoor environment. 359 results were retrieved from the database, including 13 from Scopus, 58 from Science Direct, and 284 from PubMed. There were over 3,140 results on Google Scholar, but we reviewed only the first ten pages with 6 articles related. The search string is as in Table 2. The database search results were stored in Excel and final articles were transferred to Mendeley's online reference manager.

Table 2. Search strings from Web of Science, SCOPUS and Science Direct

Search strings from SCOPUS	TITLE-ABS-KEY ((("Kindergarten" OR "Preschool") AND ("indoor air pollution" OR "indoor air quality") OR ("Particulate Matter") OR ("Ozone") OR ("Volatile Organic Compounds") OR ("Formaldehyde") OR ("Carbon Monoxide") OR ("Carbon Dioxide") AND ("Malaysia"))) AND (LIMIT-TO (PUBYEAR, 2020) OR LIMIT-TO (PUBYEAR, 2019) OR LIMIT-TO (PUBYEAR, 2016) OR LIMIT-TO (PUBYEAR, 2015) OR LIMIT-TO (PUBYEAR, 2014) OR LIMIT-TO (PUBYEAR, 2013) OR LIMIT-TO (PUBYEAR, 2012) OR LIMIT-TO (PUBYEAR, 2011)) Result = 13 articles
Search strings from Science Direct	("Kindergarten" OR "Preschool") AND ("indoor air pollution" OR "indoor air quality") OR ("Particulate Matter") OR ("Ozone") OR ("Volatile Organic Compounds") OR ("Formaldehyde") OR ("Carbon Monoxide") OR ("Carbon Dioxide") AND ("Malaysia"). Year: 2011-2020 Result = 56 articles
Search strings from	Search: ("Kindergarten" OR "Preschool") AND ("indoor air pollution" OR "indoor air quality") OR ("Particulate Matter") OR ("Ozone") OR ("Volatile Organic

PubMed (MEDLINE)	Compounds") OR ("Formaldehyde") OR ("Carbon Monoxide") OR ("Carbon Dioxide") AND ("Malaysia"). Filters: Abstract, Free full text, Full text, in the last 10 years Result = 284 articles

3.2 Inclusion/exclusion criteria

Only published research papers in English, studies conducted in Malaysia and papers with quantitative data analysis are included in this review. The year of publication only includes articles from 2011 onwards. During this year, indoor air quality research began to receive attention until 2020, which symbolizes the year of maturity of the study [30].

3.3 Paper screening and data extraction

The paper review flow process involved four stages, as shown in figure 1. After completing the searches, 359 titles and abstracts were filtered to identify studies that met the inclusion criteria, and from these 359 titles, 20 duplicate articles were removed. Authors manually checked the eligibility of the 339 articles to confirm that the articles met the criteria by reading the title and abstract. A total of 19 articles related to the kindergarten environment were included for the next process, while 320 articles were excluded since the articles focused on immunology, microbiology, medicine, genetics and molecular biology. After evaluation, 2 articles were excluded from further review as they are unrelated to the study objective.

Figure 1. Numbers of articles at each stage of the review process (PRISMA diagram)

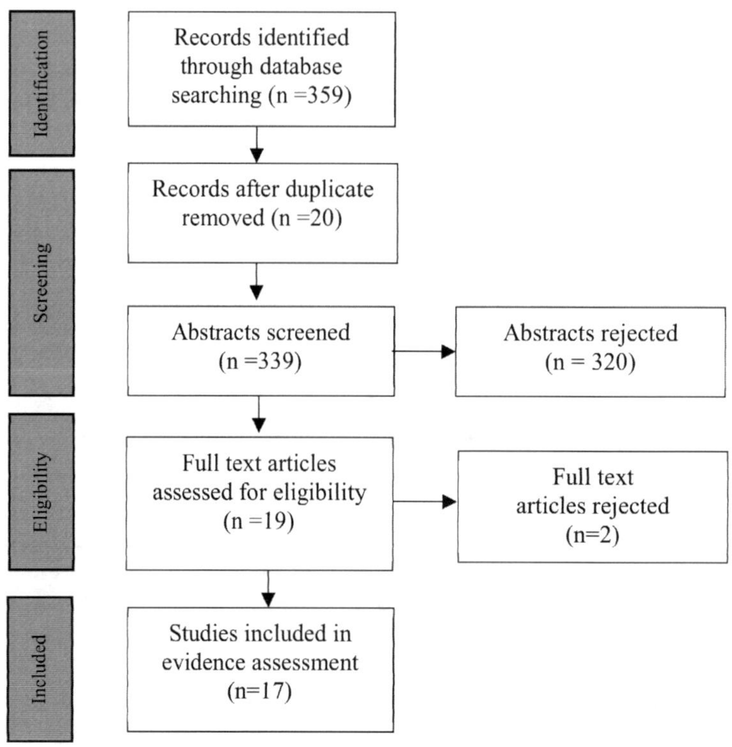

Table 3. Scope of research in articles

	Kamaruzzaman & Razak [31]	Salleh et. al [32]	Aziz et. al [33]	Rawi et. al [34]	Nor Hisyam & Juliana [35]	Choo et al. [36]	Kamaruddin et. al [37]	Wesley & Jalaluddin [38]	Salleh et. al [39]	Salleh et. al [40]	Supu & Jalaluddin [41]	Nazri et. al [42]	Sarijuddin et. al [43]	Zainudin et. al [44]	Rahman et. al [45]	Nahar & Salleh [46]	Abdullah et. al[47]
Physical Parameter																	
Indoor temperature	+	+	+	+	+	+	+	+	+	+	+	+	+	-	+	-	+
Relative humidity	+	+	+	+	+	+	+	+	+	+	+	+	+	+	+	-	+
Air Velocity	-	+	+	+	+	+	+	-	+	+	+	-	-	+	-	-	+
Indoor Chemical Pollutant																	
Volatile Organic Compound	+	+	-	+	-	+	+	+	-	+	+	-	-	+	+	-	-
Carbon Monoxide	-	+	-	+	-	+	+	-	-	+	-	-	+	-	+	-	+
Nitrogen Dioxide	-	-	+	-	+	-	-	-	-	-	-	-	-	-	+	-	-
Formaldehyde	+	+	-	-	-	-	-	-	-	+	-	-	-	-	-	-	-
Carbon Dioxide	+	+	-	+	-	+	-	-	+	+	-	-	+	-	+	+	+
Particulate Matters (PM<1)	-	-	-	-	-	-	-	-	-	-	-	-	-	+	+	-	+
Particulate Matters (PM2.5)	-	-	+	-	+	+	-	-	-	-	+	-	-	+	-	-	+
Particulate Matters (PM10)	-	+	+	+	+	+	+	+	-	+	+	+	-	+	-	-	+
Indoor Biological Pollutant																	
Fungi/Mould	-	-	-	-	-	-	-	+	-	-	+	-	-	-	-	-	-
Bacteria	-	-	-	-	-	-	-	+	-	-	+	-	-	-	-	-	-
Outdoor Chemical Pollutant																	
Volatile Organic Compound	-	+	-	-	-	-	-	-	-	+	-	-	-	-	-	-	-
Carbon Monoxide	-	+	-	-	-	-	-	-	-	+	-	-	-	-	-	-	-
Carbon Dioxide	-	+	-	-	-	-	-	-	+	+	-	-	-	-	-	-	-

(+) Factor studied (-) Factor not studies

4. Result and discussion

As indicated in Table 3 above, there were 17 articles selected in this study, with 16 case studies conducted in Selangor and one in Pahang. Overall, the studies used a cross-sectional method and case studies involving landed kindergarten (residential building and institution building) and non-landed kindergarten (commercial building). Kindergarten classrooms can be grouped into three (3) categories: 1. Fully natural ventilated classroom; 2. Fully air-conditioned classroom and 3. Mix ventilation classroom. The main indoor air problems identified in kindergarten are physical parameters (temperature and humidity), carbon dioxide, particulate matter, volatile organic compounds, carbon monoxide, bacteria and fungi.

4.1 Physical parameters

According to ICOP 2010, the recommended indoor temperature ranges from 23ºC–26ºC and the recommended indoor relative humidity (RH) is between 40%-70%. The acceptable range tends to be higher than the recommended ASHRAE Standards 55. As Malaysia is located in a tropical region, warm and humid, all natural ventilated classroom temperatures are between 28°C to 34.90°C which were outside of the temperature comfort zone. Salleh *et al.* [32] reported that the mean indoor temperatures of naturally ventilated classrooms were higher than the outdoor temperature. The temperature profile rises slightly from 9.00 am to 12.00 noon. Most air-conditioned kindergartens fell within the recommended indoor temperature range.

According to Rahman *et al.* [45], all air-conditioned classrooms showed indoor relative humidity ranging between 44.3% to 47.5% and met the ICOP 2010 standard. However, natural ventilated classrooms recorded a higher mean than recommended indoor relative humidity, ranging from 58.48 % to 77.5%. The higher percentage of indoor relative humidity detected between 7.00 am to 9.00 am and the value falling over time based on temperature [31]. Higher temperature and relative humidity levels can also raise the concentrations of some indoor pollutions. Maintaining the relative humidity level within range is crucial where excessive indoor relative humidity promotes mold growth while low percentage can cause problems such as dry eyes, nose, ears and throat.

4.2 Carbon dioxide

Indoor air quality and ventilation rate can be determined by measuring the CO_2 content in the indoor environment [26]. According to ICOP 2010, CO_2 exposure should not exceed 1000 ppm for an 8-hour period. Ten studies investigated indoor CO_2 with mean and median concentrations ranging between 537.04 ppm to 1680.35 ppm [31], [32], [34], [36], [39], [40], [43], [45]–[47]. Kindergartens in urban areas recorded high levels of CO_2 compared to kindergartens situated in suburban and rural areas [34], [36], [43]. Some studies reported high CO_2 concentrations in kindergartens classrooms exceeding 1000 ppm during class activities [31], [39], [40], [45]. Indoor CO_2 concentration increased during learning time, decreased during recess (10 am) and increased again when the classes began [46]. This situation revealed that children's respiratory activity was the primary generator of indoor CO_2. CO_2 concentrations fall during the break period as windows and doors are opened and occupancy is reduced.

The level of CO_2 in the classroom also depends on the type of ventilation used and occupant density. Furthermore, [36] found a significant connection between a high level of CO_2 and occupancy, which became even more apparent in the kindergarten that used an air-conditioning system. A study was undertaken on five kindergartens that employ different ventilation systems, with four of ten classes using air conditioners and the others using natural ventilation equipped with ceiling fans (1 piece/class). In each classroom, student capacity ranges from 19 to 21 children. All four classes showed CO_2 levels above 1000 ppm. The findings of this study coincided with Rahman *et al.* [45] as all air-conditioned classrooms in three kindergartens also recorded CO_2 levels between 1068-1090 ppm with a capacity of 16, 19 and 24 children in each kindergarten. Most kindergartens with natural ventilation systems recorded CO_2 level below 1000 ppm [34], [36], [37], [43], [47]. However, Nahar and Salleh [46] found that a high density of 19-22 students in each classroom utilizing natural ventilation also contributed to high CO_2 levels. Among contributing factors are inadequate ventilation, a small number of windows and short window opening periods. Based on the findings, it can be inferred that classrooms density that complies 2m2/person seemed to have a good IAQ, with indoor CO_2 mean value concentrations below the ICOP 2010 [46].

4.3 Particulate Matter

Particulate matter is made up of solid particles and liquid droplets that are suspended in the air. Particulate matter can be found in various shapes, sizes, origins, and chemical compositions, all of which significantly impact health. Indoor particulate matter concentrations in kindergarten classrooms was reported in twelve articles [32], [33], [34]–[38], [40], [42], [44], [47]. Of those studies, twelve (12) studies reported PM_{10}, six (6) studies reported $PM_{2.5}$, two (2) studies reported PM_1 and one (1) study reported $PM_{0.1}$. According to ICOP 2010 permissible particulate matter is 0.15mg m-3 and other national guidelines that are being referred are the new Malaysian Ambient Air Quality Standard.

Based on articles that we studied, most kindergartens that utilized natural ventilation recorded high PM_{10} and $PM_{2.5}$. Twelve studies recorded PM10 readings and four of the studies found that PM_{10} readings exceeded 100 µg m–3. A study was done by Salleh *et al.* [32] in five urban kindergartens (10 classrooms) showed that the mean concentration of indoor particulate in natural ventilation classrooms ranged from 158.57 µg/m3 to 316.24 µg/m3. Conversely, in air-conditioned classrooms, the value of indoor PM_{10} mean concentration ranged from 41.00 µg/m3 to 73.48 µg/m3 were within the recommended value. Salleh *et al.* [32] concluded that the kindergarten location between 20m to 500m from the main road and surrounded by commercial area contributed to higher indoor particulate matter.

In a cross-sectional comparative study done by Chua *et al.* [36], the location of kindergartens in urban and suburban contributed to the higher concentration of PM_{10} and $PM_{2.5}$ than rural areas of Selangor. PM_{10} concentrations in all kindergartens are below the recommended standard, urban (121.85 ± 33.45) µg/m3; suburban (119.46 ± 43.10) µg/m3 and rural (80.46 ± 24.35) µg/m3. The indoor concentration of $PM_{2.5}$ was highest in urban kindergarten with a mean of 112.62 ± 32.82 µg/m3. The mean concentration of $PM_{2.5}$ for suburban and rural kindergartens were 109.31 ± 43.30 µg/m3 and 71.77 ± 17.28 µg/m3. However, all kindergartens had mean indoor $PM_{2.5}$ above the recommended standard. PM_{10} and $PM_{2.5}$ levels in urban kindergartens were primarily caused by road traffic, emissions from neighboring industrial operations and construction work nearby.

In another study conducted by Zainudin *et al.* [44], they reported a higher reading of $PM_{0.1}$, $PM_{2.5}$, and PM_{10} at kindergartens in the urban areas with median (IQR) concentration; 90.50 ± 36.73 µg/m3, 83.10 ± 254.0 µg/m3, 86.23 ± 299.0 µg/m3 compared to a comparative group; 30.50 ± 6.00 µg/m3, 37.90 ± 9.00 µg/m3, 34.77 ± 45.00 µg/m3 respectively. Besides being close to the main road, the high concentration of indoor particulate in urban kindergarten can be explained by the daily cleaning activities performed by the kindergarten's staff and indoor combustion activities. Similarly, a study by Rawi *et al.* [34] also observed that indoor cooking activity is one of the contributing factors to higher $PM_{2.5}$. It is recommended that an exhaust fan and kitchen hood be provided to minimize cooking emissions that subsequently promote a healthy kindergarten environment.

4.4 Volatile Organic Compounds

Volatile organic compounds (VOCs) are gases that contain chemicals emitted from solids or liquids. The recommended threshold level for TVOC in ICOP 2010 is 3 ppm. Indoor VOCs concentrations in kindergarten have been reported in ten articles [31], [32], [34], [36]–[38], [40], [41], [44]-[45]. They were found in kindergarten classrooms and reported indoor VOCs ranged from non-detectable [41] to 217.13 ppb [32]. The mean of VOCs in classrooms was higher in an urban area at 0.11 ppm than in rural areas at 0.08 ppm [34]. According to Chua *et al.* [36], indoor VOCs concentration levels were highest in order of suburban 33.45 ppb, urban 29.64 ppb and rural 13.52 ppb. The findings show that VOCs are produced in the classroom regardless of whether the kindergarten is in an industry, urban or rural setting [38][44].

Salleh *et al.*[32] also found that the VOCs mean concentrations were higher in four mechanically ventilated kindergarten classrooms; 157.49 ± 41.16 ppb, 149.95 ± 49.72 ppb, 109.28 ± 36.33 ppb and 217.13 ± 25.77 ppb compared to six naturally ventilated classrooms; 105.15 ± 26.05 ppb, 51.61 ± 35.17

ppb, 23.60 ± 24.37 ppb, 36.82 ± 41.85 ppb, 18.50 ± 18.66 ppb and 95.01 ± 42.46 ppb respectively. Research by Wesley & Jalaludin [38] found no significant differences between VOCs concentration in kindergarten in urban and suburban areas. However, there was a strong link between VOCs concentration and rising levels of respiratory inflammation in kindergarten children. In another study carried out by Rawi *et al.*[34], they reported that VOCs in kindergarten classrooms were greater in urban areas (Bangi) than in rural areas (Balakong), with a significant difference between them. Furthermore, significant associations between a cough and wheezing with indoor VOCs concentration were discovered.

4.5 Carbon Monoxide

Carbon monoxide (CO) is a toxic yet fairly unreactive gas. It is a colorless, odorless, and tasteless gas created by the incomplete combustion of natural gas, diesel, or gasoline in automobile engines [48].Indoor CO concentrations in kindergarten classroom was reported in eight articles[32], [34], [36], [37], [40], [43], [45], [47]. Studies report indoor CO mainly comes from outdoor sources related to traffic. As a result, indoor concentrations are higher in kindergartens located in urban areas and naturally ventilated buildings. Studies about CO levels in kindergarten environments reported an average concentration ranging from 0.1 to 4.48 ppm. Another study found that the CO concentration is higher in urban kindergarten (4.48 ppm) than suburban (3.38 ppm) dan rural (3.04 ppm) [36]. However, Rawi *et al.* [34] recorded higher CO in rural areas (1.0 ppm) compared to urban areas (0.8ppm). The high CO content in rural kindergarten was due to proximity to busy roads and penetration of gas emissions released from traffic congestion during peak hours into kindergarten classrooms.

4.6 Fungi and Bacteria

Indoor fungi concentration was higher when there was mold/water damage in the building structure. In some people, exposure to visible mold or mold odor can cause adverse health effects such as upper respiratory tract symptoms, cough, wheeze, and asthma. The ICOP 2010 standard for total fungal count is 1000 CFU/m3 and 500 CFU/m3 for the total bacterial count. Studies investigating total indoor fungi concentrations reported results ranging from 50 to 402 CFU/m3 and bacteria concentration from 148 to 550 CFU/m3. In studying these biological contaminants, Supu and Jalaludin [41] and Wesley and Jalaludin [38] measure fungi and bacteria on a 4 – hour duration during normal class activities. Nazri *et al.* [42] took a measurement reading for 8 hours at ten stations (kindergartens, nurseries, and control stations). They discovered that the kindergarten with the highest number of children had a higher CFU count for fungi and bacteria. These studies revealed that biological contaminants would be higher if the IAQ parameters, including temperature and relative humidity, did not meet the acceptable guideline.

5. Conclusion

We reviewed the local scientific literature on indoor air pollution levels in kindergartens and extracted data from 17 studies published over ten years. The indoor kindergarten environment in Malaysia's tropical climate has been exposed to various pollutants. Our recommendation is to investigate further outdoor sources to identify the indoor/outdoor ratio. Indoor concentrations are the result of both indoor and outdoor-generated pollution. The second recommendation is to increase sampling duration from 8 hours to 24-72 hours to understand the pollution pattern of occupied and unoccupied space. Additionally, this practice is also beneficial to understand the activities and routines organized daily in kindergartens. In this current pandemic, it is critical to implement indoor air pollution mitigation measures in kindergarten space to lower levels and minimize the harmful impacts on children's respiratory health.

Acknowledgment

The authors would like to express their deepest gratitude to the Ministry of Higher Education (MOHE) Malaysia, Universiti Malaya (UM) and Universiti Teknologi MARA (UiTM). This work was supported by the Institute of Research Management and Services (IPPP), Impact-Oriented Interdisciplinary Research Grant Programme (IIRG), University of Malaya, Kuala Lumpur, Malaysia [grant number: IIRG015A-2019].

References

[1] Ritchie H and Roser M 2018 Urbanization - Our World in Data Retrieved on August 5 https://ourworldindata.org/urbanization#citation
[2] World Health Organization 2016 Ambient Air Pollution : A Global Assessment of Exposure and Burden of Disease Geneva
[3] Salthammer T *et al.*, Children's well-being at schools: Impact of climatic conditions and air pollution *Environ. Int.* **94** 196–210
[4] USEPA 2021 Reference Guide for Indoor Air Quality in Schools Retrieved on August 25 https://www.epa.gov/iaq-schools/reference-guide-indoor-air-quality-schools#Introduction
[5] Morawska L *et al.* 2017 Airborne particles in indoor environment of homes, schools, offices and aged care facilities: The main routes of exposure *Environ. Int.* **108** (April) 75–83
[6] Salvi S 2007 Health effects of ambient air pollution in children *Paediatr. Respir. Rev.* **8** (4) 275–280
[7] Matthews E and Lippman P C 2020 The Design and Evaluation of the Physical Environment of Young Children's Learning Settings, *Early Child. Educ. J.* **48** (2) 171–180
[8] Ata S, Deniz A and Akman B 2012 The Physical Environment Factors in Preschools in Terms of Environmental Psychology: A Review *Procedia - Soc. Behav. Sci.* **46** 2034–2039
[9] Inan H 2009 The third dimension in preschools: preschool environments and classroom design *Eur. J. Educ. Stud.* **1** (1) 1–66
[10] Satterlee D J, Molavi J M and Williams M E 2015 An Evaluation of Early Education Based on Physical Environmental Guidelines *SAGE Open*, **5** (2) 1–11
[11] Zhang S, Mumovic D, Stamp S, Curran K, and Cooper E 2021 , "What do we know about indoor air quality of nurseries? A review of the literature," *Build. Serv. Eng. Res. Technol.*
[12] Ministry of Education 2018 Malaysia Education Statistics 2018
[13] ASHRAE 2019 Standards 62.1 & 62.2 *Standard 62.1 & 62.2-2019 Ventilation for Acceptable Indoor Air Quality* https://www.ashrae.org/technical-resources/bookstore/standards-62-1-62-2 Retrieved on Oct 16
[14] Manisalidis I, Stavropoulou E, Stavropoulos A and Bezirtzoglou E 2020 Environmental and Health Impacts of Air Pollution: A Review *Front. Public Heal.* **8** (February,) 1–13
[15] Mandin C *et al.* 2017 Assessment of indoor air quality in office buildings across Europe – The OFFICAIR study *Sci. Total Environ.* **579** 169–178
[16] Moreno-Rangel A, Sharpe T, McGill G and Musau F 2020 Indoor air quality in passivhaus dwellings: A literature review *Int. J. Environ. Res. Public Health* **17** (13) 1–16
[17] Elbayoumi M, Ramli N A, Yusof N F F M and Al Madhoun W 2015 Seasonal Variation in Schools' Indoor Air Environments and Health Symptoms among Students in an Eastern Mediterranean Climate *Hum. Ecol. Risk Assess.* **21** (1) 184–204
[18] Challoner A and Gill L 2014 Indoor/outdoor air pollution relationships in ten commercial buildings: PM2.5 and NO2 *Build. Environ.* **80** 159–173
[19] Lee J H *et al.* 2014 Relationship between indoor air pollutant levels and residential environment in children with atopic dermatitis," *Allergy, Asthma Immunol. Res.*, **6** (6) 517–524,
[20] Mainka A and Zajusz-Zubek E 2015 Indoor air quality in urban and rural preschools in upper Silesia, Poland: Particulate matter and carbon dioxide *Int. J. Environ. Res. Public Health* **12** (7) 7697–7711
[21] Basińska M *et al.* 2019 Impact of physical and microbiological parameters on proper indoor air quality in nursery *Environ. Int.* **132** (March) 105098
[22] Błaszczyk E , Rogula-Kozłowska W, Klejnowski K, Kubiesa P, Fulara I and Mielżyńska-Švach D 2017 Indoor air quality in urban and rural kindergartens: short-term studies in Silesia, Poland, *Air Qual. Atmos. Heal.* **10** (10) 1207–1220
[23] Branco P T B S, Alvim-Ferraz M C M, Martins F G and Sousa S I V 2014 The microenvironmental modelling approach to assess children's exposure to air pollution - A

review *Environ. Res.* **135** 17–332
[24] Branco P T B S, Alvim-Ferraz M C M, Martins F G and Sousa S I V 2019 Quantifying indoor air quality determinants in urban and rural nursery and primary schools *Environ. Res.* **176** (June)
[25] Jung C C , Hsu N Y and Su H J 2019 Temporal and spatial variations in IAQ and its association with building characteristics and human activities in tropical and subtropical areas," *Build. Environ.* **163** (July)
[26] Department of Occupational Safety and Health 2010 Industry Code of Practice On Indoor Air Quality 2010
[27] Moher D *et al.* 2009 Preferred reporting items for systematic reviews and meta-analyses: The PRISMA statement *PLoS Med.* **6** (7)
[28] Okoli C 2015 A guide to conducting a standalone systematic literature review *Commun. Assoc. Inf. Syst.* **37** (1) 879–910
[29] Liberati A *et al.* 2009 The PRISMA statement for reporting systematic reviews and meta-analyses of studies that evaluate health care interventions: explanation and elaboration **62** (10)
[30] Kraus S and Dasí-rodríguez S 2020 The art of crafting a systematic literature review in entrepreneurship research *Int. Entrep. Manag. J.* **16** 1023–1042
[31] Kamaruzzaman S N and Razak R A 2011 Measuring Indoor Air Quality Performance in Malaysian Government Kindergarden *J. Build. Perform.* **2** (1) 70–79
[32] Salleh N M, Kamaruzzaman S N and Mahyuddin N 2013 Sick Building Symptoms Among Children in Private Pre-Schools in Malaysia: Association of Different Ventilation Strategies *J. Build. Perform.* **4** (1) 73–81
[33] Aziz A, Jalaludin J, and Bakar S A 2014 Indoor Air Pollutants Exposure and the Respiratory Inflammation (FeNO) among Preschool children in Hulu Langat, Selangor *Adv. Environ.* **8** (15) 164–170
[34] Rawi N A M N, Jalaludin J and Chua P C 2014 Indoor Air Quality and Respiratory Health Among Malay Preschool Children in Selangor *Biomed Res. Int.* **388** 539–547
[35] Noor Hisyam N H and Jalaludin J 2014 Association between Indoor PM10, PM2.5 and NO2 with Airway Inflammation among Preschool Children at Industrial and Sub-Urban Areas *Adv. Environ. Biol.* **8** (15) 149–159
[36] Chua P C, Jalaludin J, Hamedon T R and Adam N M 2015 Indoor air quality assessment and lung functions among children in preschool at Selangor, Malaysia *Adv. Environ. Biol.* **9** (9) 1–9
[37] Kamaruddin A S, Jalaludin, J and Chua P C 2015 Indoor air quality and its association with respiratory health among malay preschool children in Shah Alam and Hulu Langat, Selangor *Adv. Environ. Biol.* **9** (9) 17–26
[38] Wesley A D and Jalaludin J 2015 Indoor Air Pollutant Exposure and Eosinophil Cationic Protein as an Upper Airway Inflammatory Biomarker among Preschool Children in *Procedia Environmental Sciences* **30** 297–302
[39] Salleh N M, Esa M, Kamaruzzaman S N, Mahyuddin N and Darus F M 2016 Barrier of Ventilation Modes to Achieve A Balance Environment In Refurbished Preschools *J. Teknol.* **5** 261–270
[40] Salleh N M, Salim N A A, Kamaruzzaman S N, Mahyuddin N and Darus F M 2016 The Prevelence of SBS and Absenteeism among Children in Urban Refurbished Private Preshools *MATEC Web Conf.* **66** (January)
[41] Supu A and Jalaludin J 2017 Indoor Air Quality and Its Association With Respiratory Health Preschool Children in Shah Alam and Hulu Langat , Selangor *Malaysian J. Public Heal.* **1** (May) 78-88
[42] Nazri N S M , Aris M S M, and Din S A M 2017 Indoor air quality assessments in selected child care institutions (nurseries & kindergarten) at Kuantan, pahang: the implication of occupancy on iaq *J. Built Environ. Technol. Eng.* **2** 175–181
[43] Sarijuddin F A *et al.*, 2017 Assement on level of indoor air quality at kindergartens in Ampang

Jaya, Selangor, Malaysia *J. Fundam. Appl. Sci.* **9** (4S) 801–811
[44] Zainudin M A, Jalaludin J and Sopian N A 2019 Indoor Air Quality (IAQ) in Preschools and Its Association with Respiratory Inflammation among Pre-schoolers *Malaysian J. Med. Heal. Sci.*, **15** (4) 12–18
[45] Rahman M A A, Awang M, Mustafa M S S and Yusop F 2019 Evaluation and measurement of indoor air quality in the preschool building in *IOP Conf. Ser.: Earth Environ. Sci.* **373** 012018
[46] Nahar A A and Salleh N M 2019 Co_2 Field Evaluation Of Natural Ventilation Classroom In Refurbished Preschool in *The European Proceedings of Multidisciplinary Sciences* 527–535
[47] Abdullah S, Abd Hamid F F, Ismail M, Ahmed A N and Wan Mansor W N 2019 Data on Indoor Air Quality (IAQ) in kindergartens with different surrounding activities *Data Br.* **25** 103969
[48] US EPA 2021 Basic Information about Carbon Monoxide (CO) Outdoor Air Pollution Retrieved on August 5 https://www.epa.gov/co-pollution/basic-information-about-carbon-monoxide-co-outdoor-air-pollution

Air Quality Status in Konya City Centre, Konya, Turkey during Pandemic Covid-19

N Z Yahaya[1], J Jalaludin[2], H Toros[3,] and S Dursun[4]
[1]Faculty of Ocean Engineering Technology and Informatics, University Malaysia Terengganu, 21030, Malaysia.
Faculty of Medicine and Health Sciences, Universiti Putra Malaysia, Serdang, Selangor
[3]Department of Meteorology, Faculty of Aeronautics and Astronautics, Istanbul Technical University, Maslak, Istanbul, Turkey.
[4]Environmental Engineering Department, Engineering and Natural Science Faculty, Konya Technical University, Konya, Turkey.
Corresponding author: nzaitun@umt.edu.my

Abstract. High developed industry regions, provincial centers with a heavy traffic and dense populations with cold winters, using low-quality fossil fuel consumption have an effect on quality of life especially for people with respiratory diseases. The air quality data gathered from air monitoring stations for the City Center of Konya, Turkey were analysed statistically during the period when curfew due to the Covid-19 pandemic in 2020. The restrictions that cause the reduction of vehicle exhaust emissions, which are important factors in the formation of some air pollutants, are thought to be effective in improving the air quality as well as meteorological conditions are effective on the days when the air quality is analysed. It is thought that the reduction in HC, NOx, CO and PM emissions, which can be evaluated as exhaust emissions, will be an effective factor due to the restrictions of Covid-19 pandemic. Comparing between the 2019 and 2020 MCO data, we identified that most of the gases decreased with NO_2 (-24 µg/m^3), SO_2 (-24 µg/m^3), CO (-37 µg/m^3) and an increment of O_3 of +50 µg/m^3 which indicates that the MCO and restricted of movement were give an impact to air quality levels in Konya City. The increase in O_3 values were found by the existing of the sun rays in the atmosphere with the formation of O_3 during the clean air period.

1. Introduction

Rapid population growth in cities causes many environmental problems including air pollution levels. For this reason, the changes in the air quality should be examined in more detail with its sources and related factors. It is thought that the reduction in HC, PM_{10}, SO_2, NO_x, O_3 and CO emissions, which can be evaluated as exhaust emissions, will be an effective factor due to the restrictions of vehicle movements. This conditions should not be neglected; as meteorological conditions are an important parameters in air pollution levels. In addition, atmospheric ozone reduces NO_x, CO and HC compounds from the air pollutants by oxidizing them with the reactions it creates some different compounds. Investigation of [1] showed that atmospheric PM levels were increased during winter period and increasing traffic were also evaluated. [2] detected higher level of respirable size of PM capital city Tiran, because of because of heavy and clogged traffic in city centre. [3] also show similar explanation for Istanbul city. As expected, and occurred in other cities, the concentrations of PM_{10} was dramatically increased during rush hours in the morning and evening which link to human activities especially in urban environments. As well as the emission sources, many factors have effect on environmental pollution [4]. Apart from that, rapid increasing population also increases the need for food production and enhances industrial activities to groom. An analysis of the European Environment Agency (EEA) shows that 45.8 percent of the NO_x emissions in Turkey is the result of energy use and production. The second important source of NO_x is road transport, whose contribution is 35.2 percent. Furthermore, it seems that 62 percent of SO_2 emissions comes from energy production and 23.2 percent from use of energy. This means that 85.6 percent of SO_2 emissions in Turkey can be attributed to the use and production of energy.

The Covid-19 pandemic is thought to originate from China by human contact with non-domestic animals. The origin of the Covid-19 virus is thought to have spread from the wild animal market in Wuhan, China [5]-[6]. Infectious rates of diseases occur faster than in the past, and the rate of spread increases with increasing human contact. Between the beginning of January 2020 to middle of July 2020 Covid-19 infections were detected in more than 13 million people worldwide, and about 600 thousand people died. According to the [7], there have been 5,313,098 confirmed cases of Covid-19 with 48,524 deaths from 3 January 2020, 11 June 2021 in Turkey. Covid-19 infection first detected activity in Turkey was on March 15, 2020 and the measures taken in Turkey to reduce the impact of Covid-19. The causes and possible consequences of the outbreak have not yet been adequately studied. Since contact is the most important factor in infection, methods should be developed to reduce it. It has become important to understand the impact of the Covid-19 episode on the level of pollutants in this region. This link is to prevent a sustainable environment. Table 1 shows air pollutant standards for the European Union and World Health Organizations and Turkish Air Quality that is controlled by the Ministry of Environment and Urbanization of the Turkish Republic for comparisons.

Table 1. Air quality standards for the European Union and World Health Organisation.

Air Pollutants	European Union Air Quality Standards[a]	WHO Guidelines[b]	Turkish Air Quality Standards (2019)[c]
$PM_{2.5}$ (24-hr mean)	NA	25 μg/m^3	NA
PM $_{2.5}$(Annual)	Limit Value, 25 μg/m^3	10 μg/m^3	NA
PM_{10} (Hourly)	Limit Value, 50 μg/m^3	50 μg/m^3	60 μg/m^3
PM_{10} (Annual)	Limit Value, 40 μg/m^3	20 μg/m^3	NA
O_3 (8-hr mean)	Target Value, 120 μg/m^3	100 μg/m^3	NA
NO_2 (Hourly)	Limit Value, 200 μg/m^3	200 μg/m^3	250 μg/m^3
NO_2 (Annual)	Limit Value, 40 μg/m^3	40 μg/m^3	NA
SO_2 (Hourly)	Limit Value, 350 μg/m^3	500 μg/m^3 (10 Minutes mean)	NA
SO_2 (24-hour)	Limit Value, 125 μg/m^3	20 μg/m^3	125 μg/m^3
CO (1-hr daily mean)	NA	26 ppm	NA
CO (8-hr daily mean)	Limit Value, 10 mg/m^3	9 ppm	10 μg/m^3

[a]EU Air Quality Directive (2008/50/EC)
[b]WHO, 2006, Air Quality Guidelines: Global update 2005.
[c]Air pollution in Istanbul (rvo.nl) (2017)

Many recent studies have explored impacts of the Covid-19 lockdowns or Movement Control Order (MCO) on air quality. The most common approach that has been used to determine a simple statistical analysis that compares air quality before and after the lockdowns began or during the lockdowns with the same periods in previous years [8]-[9]. Study by [10] in Spain also compared the air quality before and after lockdown started for periods with similar meteorological conditions. While [11] assesses nitrogen dioxide levels by using satellite observations. Similar techniques were used by [12], [13] and [14] that analysed the level of NO_2 have also been used to estimate the reduction in column NO_2 due to the lockdowns. [15] studied the air quality status during the 2020 Malaysia Movement Control Order (MCO) due to the 2019 novel coronavirus (2019-nCoV) pandemic by comparing PM $_{2.5}$ level. The study revealed that the changes of fine particulate matter ($PM_{2.5}$) at 68 air quality monitoring stations were found that the $PM_{2.5}$ concentrations showed a high reduction of up to 58.4% during the MCO. Several red zone areas (N41 confirmed Covid-19 cases) had also reduced up to 28.3% in the $PM_{2.5}$ concentration variation. Study by [16] in Klang Valley air quality data in Malaysia indicated that there were significant reductions ($p < 0.05$) of PM $_{10}$, $PM_{2.5}$, NO_2 and CO during the MCO compared with the same periods in 2019 and 2018. The highest percentage of reduction during the MCO was recorded by NO_2 with high percentage of reduction were recorded.

2. Materials and Methods

Konya is located in central Anatolia and it has the largest surface area in all the provinces of Turkey (41 thousand km^2). Konya is geographically located between 36° 41' and 39°16' north latitudes and 31°41' and 34 ° 26' east longitudes (Figure 1). The average height from the sea level is 1016 m and the height, the city center is 1028 m from the level of the sea. As of the 2020 census, Konya population is at 2.25 million people, which constitutes 2.7% of Turkey's population of 84 million. When the comparisons are made in other provinces of Konya, Turkey's 7th largest city. In terms of area it is the largest city of Turkey. The number of people per km^2 for Konya metropolitan is 55 people. Due to the fact that the annual average rainfall in Konya city is 320 mm, it made it necessary to carry out dry farming. Because summers are hot and dry in the research area. The steppe climate, which has harsh and snowy winters, is observed. When Konya's long years of data are analysed, the seasonal condition of temperature and precipitation shows a continuous fluctuation according to the years. In the region, these fluctuations have reached extreme values, leading to the development of steppe vegetation. In terms of its land structure, vegetation has the appearance of a plain steppe. Low mountain and high mountain steppe surrounded the plain steppe [17]-[18].

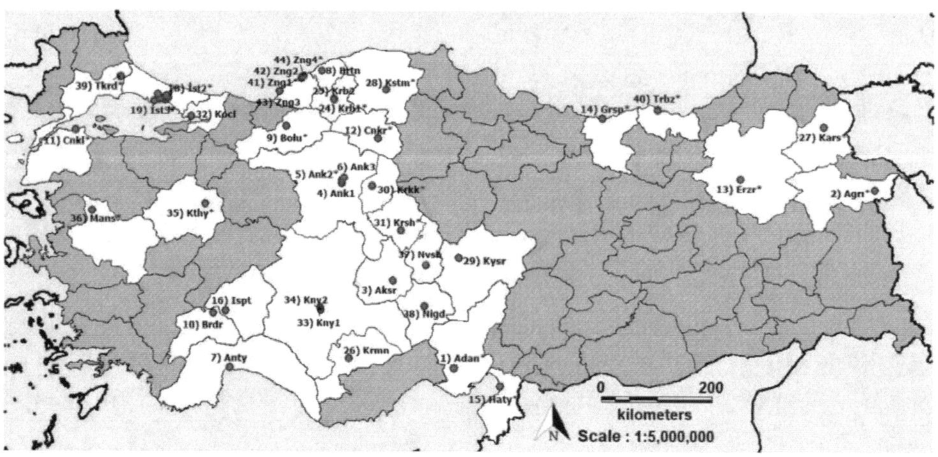

Figure 1. Location of the investigation area

Konya province has a closed basin structure and the basin consists of wide plain areas and plateaus. The Taurus Mountains cover the south of the basin and there are high mountains to the west, preventing the moist air of the Mediterranean from coming to the region. Therefore, it has a very arid climate and has an arid feature. With this feature, summer is hot and dry, and winter is cold and snowy.

Konya has less rainfall compared to most provinces in Turkey. Although it is known in Turkey as a grain warehouse development, the change of climate and water resources vary this feature. Industrial areas have developed and are developing in the city centre.

Air quality data of PM_{10}, NO_2, CO, SO_2 and O_3 were gathered from four sampling stations of Konya city, via the Ministry of Environment and Urbanization of the Turkish Republic. Data of the research area were divided in two parts that; First Period 1: before the Covid-19 measures were taken to reduce effect of pandemic problem which is between January 1 and March 15 of 2020; and Period 2: after the Covid-19 struggle started in Turkey and Covid-19 measures applied which is between March 16 and April 15 of 2020 and in June for summer time. In addition to hourly data, daily air pollutant data of 01.01.2020-15.06.2020 were compared to see the effect of Covid-19 measures and seasonal effect on pollution parameters. Same season of 2018 and 2019 were also collected and compared between 2020 and earlier data. There are various sampling methods to be used in research. According to the purpose of the research, restrictive conditions such as cost, duration and the characteristics of the audience, the most appropriate among them should be selected. Data analysis covers all the statistical methods required to summarize and evaluate the data collected within the scope of the research. In this study, R-program, an open code statistics program, was used to create the distribution maps of the data.

After the occurrence of the first infection case in Turkey, it was very quickly becoming a measure as a part of the fight against infection with various precautions. In this context, after 16th March 2020, all schools were closed, then online study started in Konya city. Furthermore, mass worship was interrupted in mosques, and quarantine measures were implemented in all metropolitan cities in Turkey. Therefore, the mobility of people has been reduced between the metropolitan cities. Furthermore, the age of 65 and under the age of 20 are forbidden to go out in all the cities of Turkey, and curfew measures were introduced for all age groups of all metropolitan areas during the weekends. And finally, commercial/touristic flights were canceled in many countries where the pandemic infection was widespread. All the restrictions have an impact on mobility and reduced pollutants that are emitted from transportation sectors.

3. Results

Investigation region where Konya metropolitan is one of the highly developed industry provinces in Turkey. Provincial living centres also have heavy traffic around settlement roads, and the settlements region has a very cold winter season. When low-quality fossil fuels consumption affects quality of life especially in people with respiratory diseases because of emission of incineration. Air pollution is increasing some days due to specific climatic conditions occurring via the topographic structure of the region. Climatic features of a region are as effective as increasing emission locally in the formation of air pollution. Especially in the winter months, there is an increase in the emission levels as the temperature decreases with increasing fuel usage for heating systems. However, some days of winter period, the negativity caused by meteorological conditions causes lower than expected levels of air quality.

4. Descriptive Statistics

Hourly particles, gases, and meteorological data were gathered from the four sampling stations of Konya city, via the Ministry of Environment and Urbanization of the Turkish Republic. All the data measurements were averaged by using a similar averaging time of 1 hour for comparison purposes. Data were gathered and separated by two datasets from 01 January 2020 to 31 December 2020 and were analysed by using a comprehensive packages R Software version 4.1.0 [19] and an Openair packages [20]-[21] and its packages statistically and graphically are summarized in Table 2. Data were also screening processes that involved the deletion of several spur data. Every outlier and extreme value were analysed by time series plot and the spur value was detected and deleted before further analyses were conducted.

5. Temporal Variation of Pollutants

Konya metropolitan is one of the highly developed industry provinces in Turkey. Provincial living centres also have heavy traffic around settlement roads. The settlements region has a very cold winter season that has some effects on concentrations of air pollution here. Other factors such as the topographic structure of the region also influenced the specific climatic condition and surface terrain. Climatic features of a region are as effective as increasing emission locally in the formation of air pollution. Especially in the winter months, there is an increase in the emission levels as the temperature decreases with increasing fuel usage for heating systems. However, some days of winter period, the negativity caused by meteorological conditions causes lower than expected levels of air quality. Four air quality parameters of Konya Erenkoy sampling station are SO_2, CO, NO_2 and O_3 given following figures. They show that SO_2, NO_2 and CO values significantly decreased and the opposite trend for O_3 values which were increased, during the period of Covid-19 outbreak, after March 15, 2020 (Figures 3-6). Figure 3 shows the SO_2 concentrations were higher in January and lowered down to10 µg/m³ and after covid-19 outbreak decreased about 5 µg/m³ then turned to normal levels as shown in Table 2. Four air quality parameters of Konya Selcuklu Municipality (Selcuklu Belediye) sampling station are PM10, SO2, CO and NO_2, there is no O_3 measurement in this station. Results of measurements for 4 pollutant parameters show that PM_{10} and NO_2 were initially lowered, but other two parameters, SO_2, and CO values were not significantly decreased.

Table 2. The summary of descriptive of pollutants during the year 2020.

Stations	EU Standard (µg/m³)	Selcuklu (mean,max/sd)	Karatay (mean,max/sd)	Meram (mean,max/sd)	Beledye (mean,max/sd)
PM_{10} (1-hr mean)	50	37.94/678.56/	37.41/523.6/39.2	31.94/297.5/30 .3	35.49/387.5/37
O_3 8-hr mean	120	NA	11.9/38.7/14.9	NA	NA
NO_2 (1-hr mean)	200	50.00/232/31.3	25.0/92.2/15.8	NA	37.29/165.1/22
SO_2 (1-hr mean)	350	17.1/257.6/21.6	11.3/89.6/9	11.55/317.4/16	10.6/127.80/9
CO (24-hr mean)	10000	780.9/9315/792	691.8/15614/1134	na	704/8876/849

Figure 2-6 shows that PM_{10} and other pollutant levels of four air quality sampling stations were significantly lowered by Covid-19 outbreak period and were increased with Sahara dust affecting periods which is common conditions in this region. Results of atmospheric PM_{10} measurement not only affected the emissions source, but also affected meteorological factors and special situations. Figure 2 shows the PM_{10} values of four different air quality sampling stations in Konya city centre during the first half of 2020 including the period of Covid-19 outbreak. Figure 2 shows that PM_{10} levels of four air quality sampling stations were significantly lowered by Covid-19 outbreak period and were increased with Sahara dust affecting periods then lowered to normal levels again. Results of atmospheric PM_{10} measurement are not only affected by the emissions source, but also affected meteorological factors and special situations. Figure 2- 4 (a and b) illustrates the time series plots for the 1-hour averaged data for PM_{10} for all two stations.

IOP Conf. Series: Earth and Environmental Science 1013 (2022) 012006 doi:10.1088/1755-1315/1013/1/012006

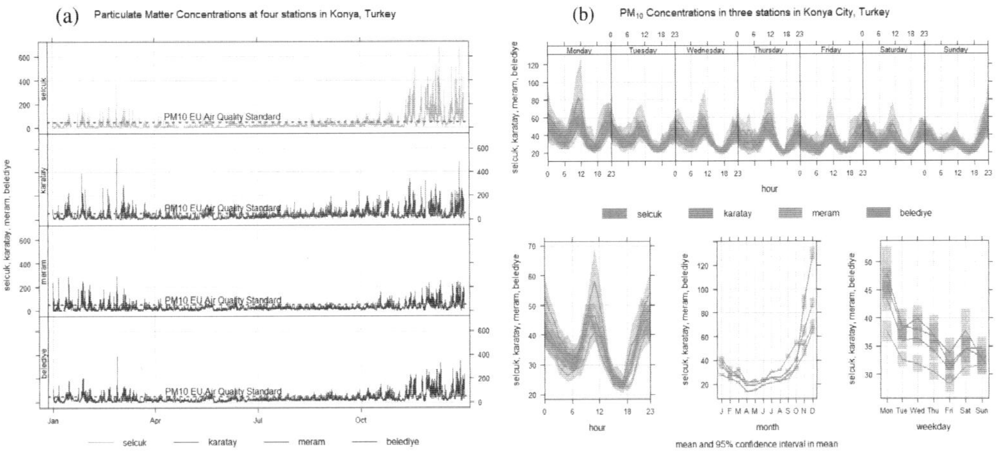

Figure 2. (a) TimePlot and (b) Time variation plots for PM_{10} concentrations at four stations in Konya City, Turkey.

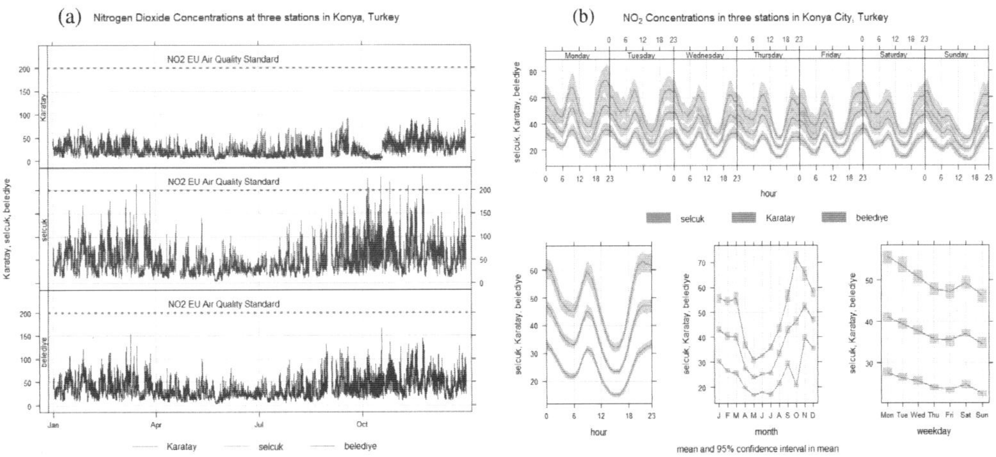

Figure 3. (a) TimePlot and (b) Time variation plots for nitrogen dioxides (NO_2) concentrations at three stations in Konya City, Turkey.

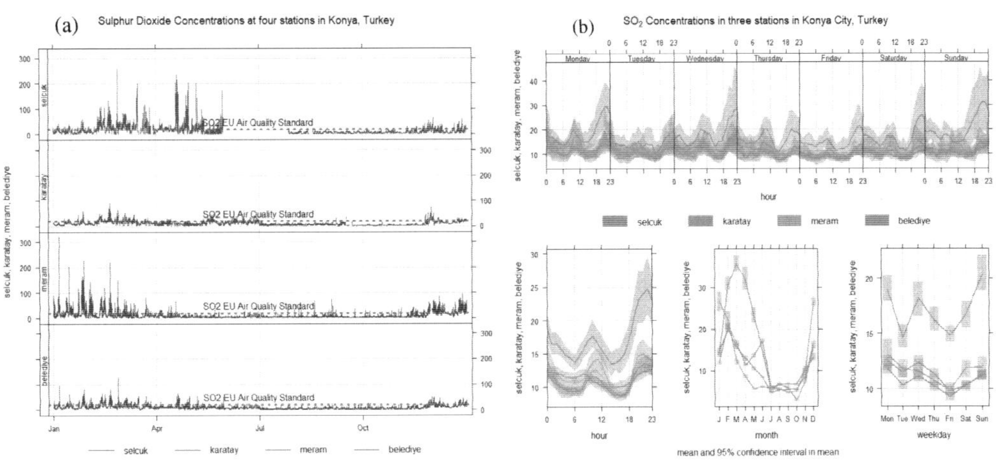

Figure 4. (a) TimePlot and (b)Time variation plots for sulphur dioxides (SO_2) concentrations at four stations in Konya City, Turkey.

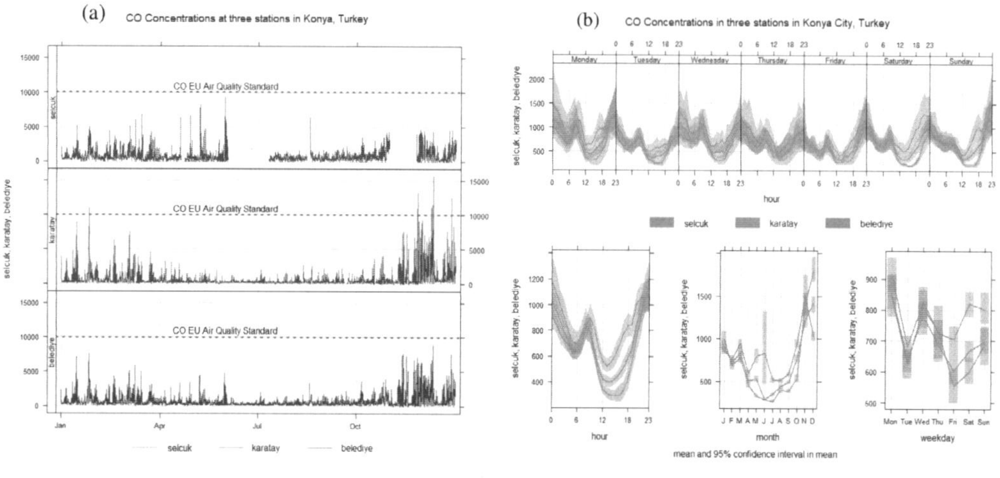

Figure 5. (a) TimePlot and (b) Time variation plots for carbon monoxide (CO) concentrations at three stations in Konya City, Turkey.

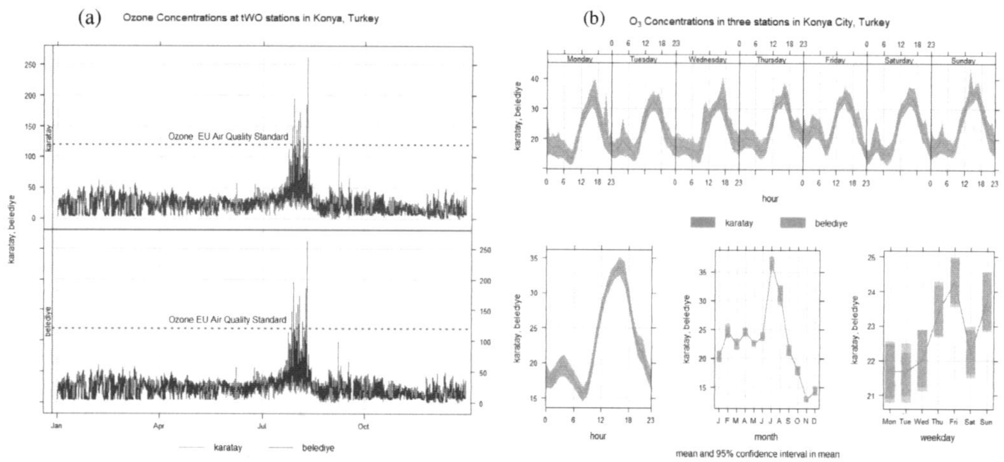

Figure 6. (a) TimePlot and (b)Time Variation plots for ozone(O_3) concentrations at two stations in Konya City, Turkey.

Figure 7 shows the PM_{10} values of three air quality sampling stations Selcuklu, Karatay and Selcuklu Belediye in Konya city centre during first half period of 2020 year. It can be seen slightly lowering from 2018 to 2019, but lowering rate is more in 2020. It is clear that reduction of PM_{10} concentrations was lower during Covid-19 outbreak period (01 January 2018, 15 June 2020) than before the outbreak and 2018 and 2019 years at same period.

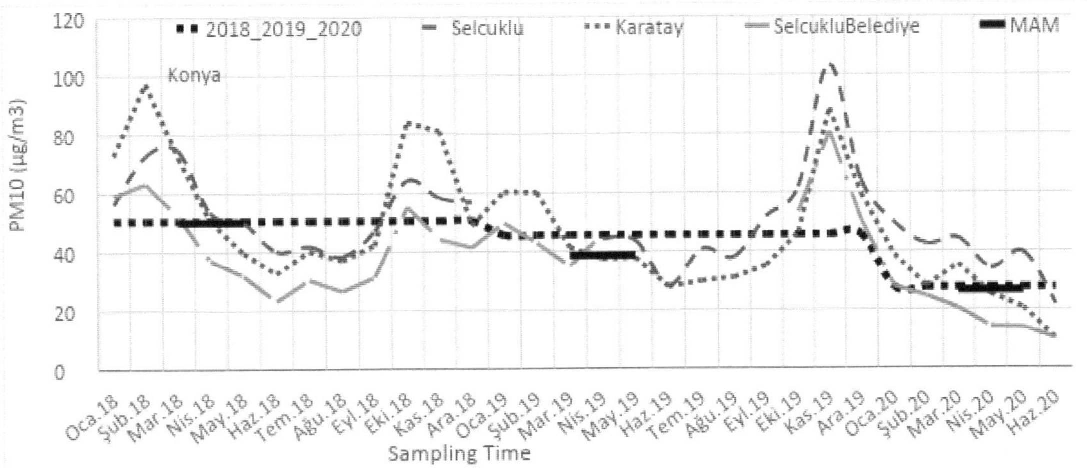

Figure 7. PM_{10} values of three air quality sampling stations Selcuklu, Karatay and Selcuklu Belediye in Konya city centre during first half period of 2020 year. Solid line (▬MAM) shows a Covid-19 outbreak period and the other two are the same period in 2018 and 2019 years respectively. Dashed line (-----) mean of each year 2018, 2019 and 2020.

It is seen that the air quality increases covid-19 outbreak period (01 January 2018, 15 June 2020) and then returns to normal levels during the period when curfew is restricted for measures taken due to the covid-19 outbreak. The restrictions that cause the reduction of vehicle exhaust emissions, which are important factors in the formation of some air pollutants, are thought to be effective in improving the air quality. It is clear that there is a significant decrement of SO_2, NO_2 CO values decreased but O_3 values were increased (Tables 3-6). Minus values (-) show decrease and plus (+) values are increased. In this study, we estimate the impact of COVID-19 on air pollution by using Eq. (1) given by [16];

Percentage of Difference (%) or PD = ((A – B) / B) x 100. Eq. (1)

Where;
A = Concentration of air pollution recorded during the MCO in 2020
B = The concentration of air pollution recorded in the same period in 2019 or 2018

The Percentage of Difference (%) or PD value of each pollutant, namely SO_2, NO_2 CO, O_3 and PM_{10}, were computed for Konya air quality data by comparing 2020 data with 2018 and 2019 data, as shown in Tables 3-7. Minus values (-) show decrement of concentrations, and (+) values were an increment of pollutants. In general, SO_2 and NO_2 and PM_{10} concentrations decrease with -12.3%, -14% and 2-4%, respectively. Meanwhile, O_3 and CO show an increment of percentage with +33% and + 21%, respectively, for both pollutants. The same phenomena have been reported previously, such as by [16] in Malaysia and by [22] effects of the season on the percentage of different pollutant levels vary between pollutants that may need further investigations and other techniques. However, O_3 pollutants are shown consistently to increase in all seasons.

Table 3. The effect of COVID-19 outbreak measures on SO_2 concentration (μg/m^3) for Konya city center.

Period	2018	2019	2020	PD (2020:2019) (%)	PD (2020:2018) (%)
Period 1 (winter)	21	14	15	5	-29
Period 2 (spring)	7	8	7	-15	5
Period 3 (summer)	6	10	7	-27	35

Table 4. The effect of COVID-19 outbreak measures on NO_2 concentration (μg/m^3) for Konya city center.

Period	2018	2019	2020	PD (2020:2019) (%)	PD (2020:2018) (%)
Period 1 (winter)	NA	59	42	-28	NA
Period 2 (spring)	19	38	31	-18	65
Period 3 (summer)	11	29	31	4	181

Table 5. The effect of COVID-19 outbreak measures on CO concentration (μg/m^3) for Konya city center.

Period	2018	2019	2020	PD (2020:2019) (%)	PD (2020:2018) (%)
Period 1 (winter)	1462	910	866	-5	-41
Period 2 (spring)	841	409	510	24	-39
Period 3 (summer)	638	269	393	46	-38

Table 6. The effect of COVID-19 outbreak measures on O_3 concentration (μg/m^3) for Konya city center.

Period	2018	2019	2020	PD (2020:2019) (%)	PD (2020:2018) (%)
Period 1 (winter)	NA	34	37	6	NA
Period 2 (spring)	NA	35	54	52	NA
Period 3 (summer)	NA	38	57	52	NA

Table 7. The effect of COVID-19 outbreak measures on PM_{10} concentration (μg/m^3) for Konya city center.

Period	2018	2019	2020	PD (2020:2019) (%)	PD (2020:2018) (%)
Period 1 (winter)	72	59	37	-37	-49
Period 2 (spring)	47	40	23	-43	-51
Period 3 (summer)	30	26	27	7	-10

6. Conclusion

During the Covid-19 pandemic quarantine, the restriction of movement of motor vehicles, transportations and closure of industries greatly influenced the concentration of air pollutants. Some measures taken during the Covid-19 pandemic period may turn into permanent habits over time. Comparing between the 2019 and 2020 MCO data, we identified that most of the gases decreased with NO_2 (-24 μg/m^3), SO_2 (-24 μg/m^3), CO (-37 μg/m^3) and an increment of O_3 of +50 μg/m^3 which indicates that the MCO and restricted of movement were give an impact to air quality levels in Konya City. Studying the characteristics of PM_{10}, CO and O_3 was important for improving the quality of life, physical health and the living environment of inhabitants. However, the factors that influenced these phenomena are still in study in most cities and expanded. The Covid-19 outbreak may be a turning point in combating global climate change as the behaviors of traveling and people's movement are restricted. The improvement in air quality can be considered as the most important gain after the pandemic. Air pollution and climate change have long term effects through complex interactions in the atmosphere and ground level surface of pollutants especially in urban environments. Understanding the spatial variability of pollutant concentrations, especially when looking into before and after MCO, will not only add to our knowledge of the mechanism of air pollution but also provide a scientific reference for implementing targeted control measures. For sustainable environmental and air quality, gains from measures should be continued. With the end of the measures, as in many parts of the world, environmental quality decreases.

7. Acknowledgements

The authors wish to thank the Ministry of Environment and Urbanization of the Turkish Republic for producing the data, Atmospheric, Air Quality and Climate Change Research Group, University Malaysia Terengganu, The ITU, Konya Technical University, University Putra Malaysia and Ministry of Environment and Urbanization of Turkish Republic, Turkey and The Clean Air Forum Society of Malaysia for the support.

References

[1] Dursun S 2019 Ambient particle matter pollution of Bosnia District of Konya City, Turkey. *Int. J. Ecosyst. Ecol. Sci.* **9** (4): 591-602

[2] Mankolli H and Lika M 2011 Some air pollution indicators in city of Tirana, Albania. Sukru Dursun, *International Journal of Global Warming* **3** pp 30–38

[3] Erdun H, Öztürk, A. Çapraz, Ö, Dursun A D, Dursun Ş and Toros H 2015 Spatial variation of PM_{10} in Turkey *Atmosfer Bilimleri Sempozyumu* **28** pp. 311-323

[4] Ayturan Z C, Çay T and Dursun S 2016 Konya City urban sprawl and environmental pollution relation *International Journal of Ecosystems and Ecology Science (IJEES)* **6**, 4

[5] World Health Organization (WHO). Retrieved July, 2020 https://www.who.int/docs/default-source/coronaviruse/situation-reports/20200423-sitrep-94-covid-19.pdf

[6] Shah A U M, Safri S N A, Thevadas R, Noordin N K, Rahman A A, Sekawi Z, Ideris A and Sultan M T H 2020 COVID-19 outbreak in Malaysia: Actions taken by the Malaysian government. *International Journal of Infectious Diseases* **97** pp 108-116

[7] World health Organization (WHO) 2021 https://covid19.who.int/region/euro/country/tr Retrieved 12 June 2021

[8] Sicard P, Marco A D, Agathokleous E, Feng Z, Xu X, Paoletti E, Rodriguez J J D and Calatayud V 2020 Amplified ozone pollution in cities during the COVID-19 lockdown. *Sci. Total Environ.* **735** 139542

[9] Sharma S, Zhang M, Anshika, Gao J, Zhang H, Kota S H 2020 Effect of restricted emissions during COVID-19 on air quality in India. *Sci. Total Environ.* **728**, 138878

[10] Tobías A, Carnerero C, Reche C, Massagué J, Via M, Minguillón M C, Alastuey A and Querol X 2020 Changes in air quality during the lockdown in Barcelona (Spain) one month into the SARS-CoV-2 epidemic. *Sci. Total Environ.* **726** 138540

[11] Ogen Y 2020 Assessing nitrogen dioxide (NO_2) levels as a contributing factor to coronavirus (COVID-19) fatality. *Sci. Total Environ.* **726**, 138605

[12] Venter Z S, Aunan K, Chowdhury S and Lelieveld J 2020 COVID-19 lockdowns cause global air pollution declines. *Proc. Natl. Acad. Sci.* **117**, 18984–18990

[13] Zhang R, Zhang Y, Lin H, Feng X, Fu T-M and Wang Y 2020 NO_x emission reduction and recovery during COVID-19 in east China. *Atmos.* **11** 433

[14] Liu F, Page A, Strode S A, Yoshida Y, Choi S, Zheng B, Lamsal L N, Li C, Krotkov N A, Eskes H, van der A R, Veefkind P, Levelt P F, Hauser O P and Joiner J 2020 Abrupt decline in tropospheric nitrogen dioxide over China after the outbreak of COVID-19. *Sci. Adv.* **6** eabc2992

[15] Samsuri A, Mansor A A, Napi N N L M, Mansor W N W, Ahmed A N, Ismail M and Ramly Z T A 2020 Air quality status during 2020 Malaysia Movement Control Order (MCO) due to 2019 novel coronavirus (2019-nCoV) pandemic. *Sci. Total Environ.* 729: 139022

[16] Latif M T, Dominick D, Hawari N S S L, Mohtar A A A and Othman M 2020 The concentration of major air pollutants during the movement control order due to the COVID-19 pandemic in the Klang Valley, Malaysia. *Sustain. Cities Soc.* **66**: 102660

[17] Kaya B and Aladağ C 2009 Analyse of Distribution Areas in Turkey and Ecologic Characteristics of Maquis and Garriques Communities. *Selcuk Univesity Sosial Sciences Institue Journal*, **22** 67-80

[18] Aka A S 2007 Konya Şehrinin Kuruluş ve Gelişimini Etkileyen Coğrafi Faktörler, PhD. Thesis, Selcuk Un. Konya, Turkey. pp 285

[19] R Development Core Team 2008 *R: A language and environment for statistical computing*. Austria: R foundation for statistical computing Vienna

[20] Carslaw D C and Ropkins K 2012 Openair - An R package for air quality data analysis. *Environmental Modelling & Software*. **27-28**, pp 52–61

[21] Carslaw D C 2019 The openair manual - open-source tools for analysing air pollution data. Manual for version 2.6-6, University of York and Ricardo Energy & Environment

[22] Said M, Zhiwen L and Tim D 2021 Comparing different approaches for assessing the impact of COVID-19 lockdown on urban air quality in Reading, UK. *Atmospheric Research* **261**: 105730

Air quality status before, during and after the pandemic COVID-19 Movement Control Order (MCO) at urban and suburban areas in Malaysia

A Z M Zahid[1*], A A A Bakar[1], N F M Halim[1] and M Z M Salleh[1]

School of Civil Engineering, College of Engineering, Universiti Teknologi MARA Johor Branch, Pasir Gudang Campus, 81750 Masai, Johor Darul Ta'zim, Malaysia
*Corresponding author: azyan8410@uitm.edu.my

Abstract. A severe acute respiratory syndrome, COVID-19 outbreak started in December 2019 reported in the city of Wuhan, China has spread rapidly in other countries around the world. Declared as a pandemic, the lockdowns, industrial shutdowns, closing international borders and overall reduced mobility of the population either for tourism or work commutes have led to air pollution reduction. The lockdown implemented worldwide including Malaysia has come as a respite to the environment, especially in air pollutant levels. Utilizing the Air Pollutant Index (API) data recorded on an hourly basis throughout different series of national lockdown phases, this study is undertaken to evaluate the air quality status before, during and after the implementation of MCO by comparing the selected urban and suburban areas in Malaysia. Six (6) locations including Bandaraya Melaka, Cheras and Alor Setar (urban) and Seri Manjung, Kangar and Kuantan (suburban) were chosen in this study. The result revealed that the air quality has shown improvements after a further extension of lockdown days at both urban (API 36.3 – 61.5%) and suburban areas (API 46.5 – 48.6%). It has indirectly changed the air quality status from 'moderate' to 'good' in all 6 urban and suburban areas. The air quality trends, however, gradually deteriorate and has increased as more restrictions were eased at a later stage. The implementation of lockdown measures not only restricted the spread of infection rate but has reduced ambient air pollutants levels that lead to enhanced air quality. The results obtained in this study can be a useful insight for the regulatory agencies in controlling air pollution and restore the environment quality.

1. Introduction

The coronavirus COVID-19 outbreak was reported to emerge in late December 2019 in Wuhan, Hubei Province, China and affects the human respiratory system. The rapid spread of this novel virus has led the World Health Organization (WHO) to declare COVID-19 as a pandemic and subsequently led the world to lockdown. To limit the continuous spread of the virus, many countries took unprecedented measures to restrict their economic activities and non-essential travelling. Malaysia recorded the earliest COVID-19 case in late January 2020 on travellers from China arriving in Johor via Singapore [1,2] The massive spikes began to occur in early March 2020 and have urged the Malaysian government to take several measures to contain the outbreak.

A nationwide Movement Control Order (MCO) was announced by Malaysia's Prime Minister on 16 March 2020 to mitigate the escalation of COVID-19 by isolating the source of the outbreak. The implementation of the first phase of MCO (MCO I) was between 18 March until 14 April 2020 [3,4]. Except for essential services, almost all businesses were not allowed to open during MCO [5]. However, as the rate of new cases per day remained consistently high, further extension of MCO (MCO II) was announced until 28 April 2020 [6]. On 23 April 2020, the MCO (MCO III) was extended by an additional two weeks, until 12 May 2020 [7]. A decline in daily infection cases observed during this phase has gradually eased the restriction, and Conditional Movement Control Order (CMCO) was implemented on 4 May 2020 in all states, except Kedah, Pahang, Kelantan, Sarawak and Sabah that continued under MCO (MCO IV) [8]. The Recovery Movement Control Order (RMCO) that was implemented on 10 June 2020 was planned to end on 31 August 2020 [9]. However, due to the continuous detection of imported cases and new clusters, RMCO was extended until the end of the year [10]. Figure 1 shows the MCO-CMCO-RMCO timeline at 13 states and 3 federal territories of Malaysia throughout 2020.

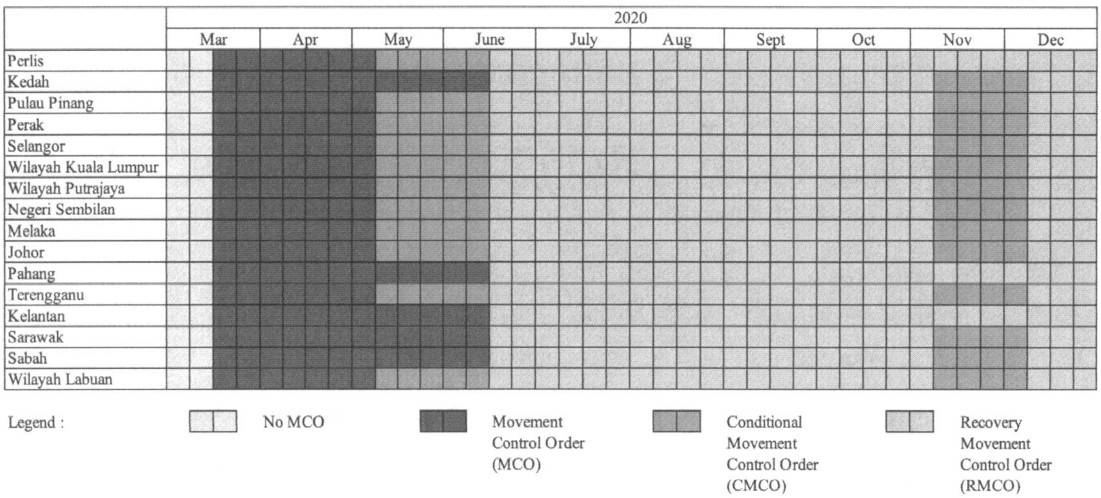

Figure 1. MCO-CMCO-RMCO timeline of Malaysia (2020)

The different implementation of lockdown phases in Malaysia was based on subsequent relaxation in the norms to restart economic and social activities. During CMCO, almost all businesses were allowed to operate. Certain outdoor sports activities were also allowed while practising social distancing. However, except for work purposes, interstate travel, as well as any events involving large gatherings, religious or assembly activities in houses of worship, were prohibited. Under the RMCO, more businesses and activities were allowed to resume operations. That includes many including hotels, gated communities, private pools, fitness centres, spas, cinemas, massage centres, meetings, weddings and religious gatherings. Interstate travel was also allowed under RMCO, except for the areas remaining under Enhanced Movement Control Order (EMCO) [5].

The sources of air pollution in Malaysia are derived from vehicles motor vehicles, industries and open combustion [11–13]. During the lockdown, the traffic and industrial emissions have reduced since some industries are suspended and people began to work from home. However, as more restrictions are uplifted and businesses and activities are resumed to operate, it affects Air Pollutant Index (API) readings recorded during the different series of national MCO, CMCO and RMCO implementation. The API, an ambient air quality measurement based on six (6) criteria pollutants used in Malaysia reports the quality of air in ranges of values (Table 1) [14]. It also reflects its effect on human health ranging from good to hazardous as stated in Table 1. In this study, the researchers will evaluate the air quality status before, during and after the implementation of MCO by comparing the selected urban and suburban areas in Malaysia. This objective interlinked the air pollution, a pressing sustainability concern with SDG 3.9 (substantial reduction of health impacts from hazardous substances) and SDG 11.6 (reduction of adverse impacts of cities on people) from the Sustainable Development Goals (SDG).

Table 1. Air quality status based on Air Pollutant Index (API) [14]

API	Air Quality Status	Health Implications
0 to 50	Good	Low pollution without any bad effect on health
51 to 100	Moderate	Moderate pollution that does not pose any bad effect on health
101 to 200	Unhealthy	Worsen the health condition of high-risk people. High-risk people = people with heart and lung complications
201 to 300	Very unhealthy	Worsen the health condition and low tolerance of physical exercises to people with heart and lung complications. Affect public health
301 and above	Hazardous	Hazardous to high-risk people and public health

2. Methods

Figure 2. Location of Continuous Air Quality Monitoring (CAQM) station at six (6) study areas in Peninsular Malaysia; Urban – Bandaraya Melaka, Cheras and Alor Setar; Suburban – Seri Manjung, Kangar and Kuantan

Six (6) locations involving three (3) urban and three (3) suburban areas in Peninsular Malaysia were chosen in this study. The selected areas include Bandaraya Melaka, Cheras and Alor Setar (urban) and Seri Manjung, Kangar and Kuantan (suburban) as illustrated in Figure 2. The identification and selection of urban and suburban areas were based on the location of the Continuous Air Quality Monitoring (CAQM) station at these areas (available at http://www.doe.gov.my/portalv1/wp-content/uploads/2013/01/Air-Qulaity.pdf). The Air Pollutant Index (API) data on an hourly basis were obtained from the website of Air Pollutant Index of Malaysia (available at http://apims.doe.gov.my/public_v2/home.html). The data covered the air quality status before MCO (1 – 17 March 2020) (n = 408), MCO I (18 March – 14 April 2020) (n = 672), MCO II (15 – 28 April 2020) (n = 336), MCO III (29 April – 3 May 2020) (n = 120), CMCO (4 May – 9 June 2020) (n = 888) and RMCO (10 June – 31 August 2020) (n = 1,992). A total of 26,496 data were used for all selected locations, while the missing data were omitted in this study.

3. Results and Discussion

Figure 3 shows the box and whisker plots of daily average API before (No MCO), during (MCO I, MCO II, MCO III and MCO IV for Kedah and Pahang) and after MCO (CMCO and/or RMCO) for urban areas and suburban areas. The bottom and top of each boxplot represent the 25th percentile (lower quartile) and 75th percentile (upper quartile) respectively. Meanwhile, the lower whisker and upper whisker represent the minimum and maximum values of the API data set in this study. In the urban area, Bandaraya Melaka recorded 51.3 of average API before MCO, 47.9 during MCO I, 44.2 during MCO II and 34.4 during MCO III. The API level however increased during CMCO (50.4) and RMCO (52.2). Meanwhile, Cheras recorded 58.2 of average API before the implementation of MCO, and further reduced to 52.2 (MCO I), 46.4 (MCO II) and 39.7 (MCO III) of average API. Once the country entering conditional and recovery MCO (CMCO and RMCO), the reading started to increase to 54.0 and 56.1, respectively. Alor Setar recorded 56.3 of average API before MCO, 51.9 during MCO I, 37.5 during MCO II and 26.5 during MCO III. Although Kedah was announced to extend the lockdown period due to the increased number of COVID-19 cases, a small increment of average API level was recorded during MCO IV (37.8) in Alor Setar. Entering RMCO, Alor Setar recorded 50.9 of the average API level.

In represents the suburban area, Seri Manjung recorded an average API of 57.0 before the implementation of MCO, 54.1 during MCO I, 50.8 during MCO II and 30.8 during MCO III. The ease of lockdown during CMCO recorded an average API level of 47.0 and has increased to 58.5 during the recovery phase (RMCO). Meanwhile, Kangar recorded 49.3 before MCO, and slowly reduced to 48.4 during MCO I and 40.0 during MCO II. MCO III recorded a 28.6 average API level before increased to 36.5 during CMCO and 45.1 after entering RMCO. Kuantan recorded a low average API level from the beginning. Before the implementation of MCO, Kuantan recorded an average API of 41.8 and reduced it to 36.0 during MCO I. MCO II recorded an average API reading of 35.5 while MCO III recorded as low as 24.0. Like Kedah, Pahang has extended the lockdown days (MCO IV) and has recorded 38.8 of the average API level in Kuantan. The recovery phase in Pahang has recorded API reading in Kuantan with 43.9.

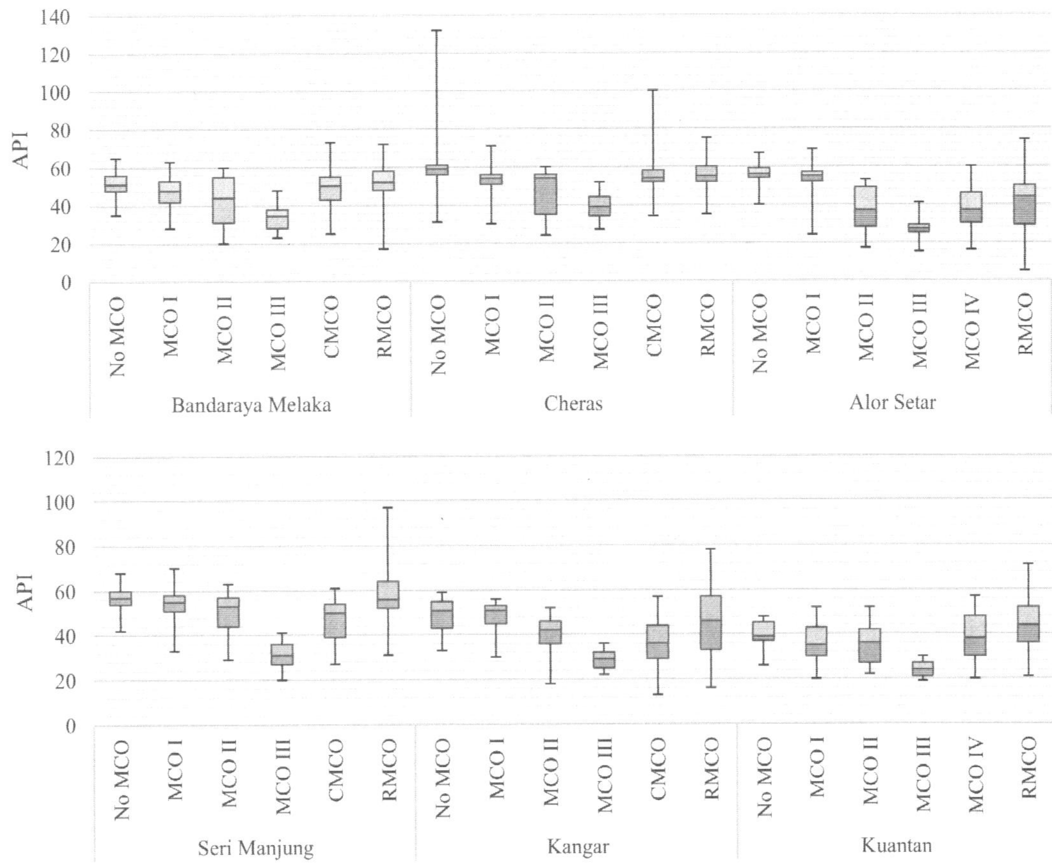

Figure 3. Box and whisker plots of daily average API before (No MCO), during (MCO I, MCO II, MCO III and MCO IV for Kedah and Pahang) and after MCO (CMCO and/or RMCO) for urban areas (top) and suburban areas (bottom)

The linear regression analysis between the daily API and the growing number of lockdown days exhibited a negative correlation at all six locations in both urban (Figure 4) and suburban (Figure 5) areas in this study. The graphs, however, show positive correlations once the CMCO was announced, except for Alor Setar (urban area) and Kangar (suburban area). With Malaysia entering into the RMCO, the API level exhibited a positive correlation further recovery phase until August 31st, 2020. Bandaraya Melaka shows a decrement in API (-6.6%) once the MCO was announced and implemented. Further reduction was recorded after the second and third extensions, MCO II (-7.7%) and MCO III (-22.2%). The ease of lockdown restrictions under CMCO starting on May 4th shows an increment of 46.5% from MCO III and further increased another 3.6% during RMCO. In Cheras, the implementation of MCO has reduced the API level by 8.5% and remained constant during MCO II. The level however decreased by 27.8% during MCO III. The relaxation of regulation has spiked the API value in the Cheras area by 38.5% during CMCO and another 1.9% during RMCO.

Alor Setar recorded a decrement only of 1.8% during MCO I. However, drastic reductions of API were recorded value during MCO II (-32.7%) and MCO III (-27%), giving the highest reduction of API level (-

61.5%) throughout full lockdown days, MCO, compared to Bandaraya Melaka and Cheras. Due to the mixed reactions among the state governments, Kedah was remained under MCO (MCO IV) to secure the positive development of the pandemic. Despite this decision, the API level in Alor Setar has increased by 37% during this phase, and further increased by 45.9% during the implementation of RMCO. Comparing urban areas in this study, the total decrement of API level during MCO I, MCO II and MCO III varies between 36.3 – 61.5%. Cheras recorded the lowest decrement of API level (-36.3%) while Melaka recorded a 36.5% reduction of API level throughout MCO implementation.

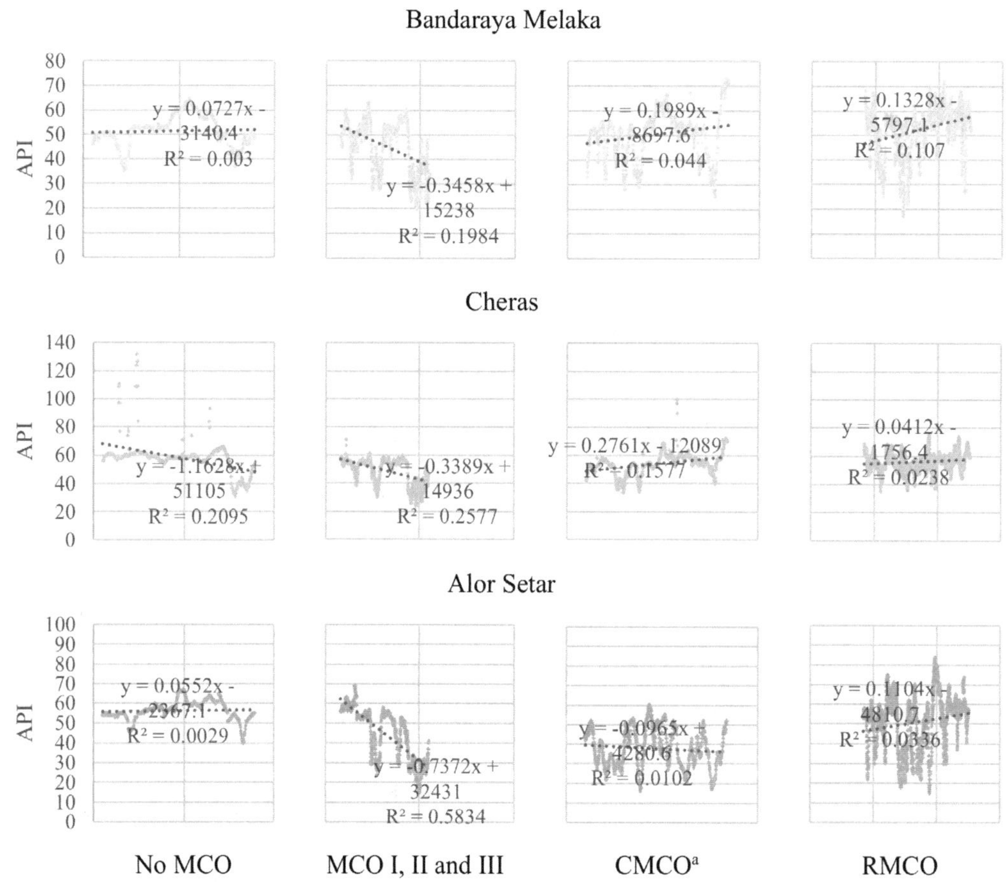

[a] except Alor Setar under MCO IV

Figure 4. Linear regression of API for urban areas; Bandaraya Melaka (top), Cheras (middle) and Alor Setar (bottom) before MCO, during MCO and after MCO (CMCO and RMCO)

In the suburban area, Seri Manjung recorded a decrement of 3.5% compared to before MCO was implemented. MCO II recorded another small reduction (-3.6%) of API level in Seri Manjung and further reduced by 41.5% during MCO III. The ease of restriction, however, has increased the API level by 61.3% and 12% during CMCO and RMCO, respectively. Meanwhile, in Kangar, no changes in API before and after the MCO I has been implemented. Nevertheless, the further extension of MCO to the second and third phases has decreased the API level by 17.6% and 31%, respectively. An increment of API levels was recorded by 24.1% during CMCO and 27.8% more during RMCO.

Kuantan, on the other hand, recorded a low API level (39) among other study areas before the implementation of MCO. The lockdown has reduced the API level further by 10.3% (MCO I). MCO II, however, recorded a small increment of API level (2.9%) and reduced further by 33.3% during MCO III. Similar to Kedah, Pahang also decided to not implement CMCO by May 4th, and continue the MCO IV along with other states, Kelantan, Sabah and Sarawak. Although remained under MCO, Kuantan recorded a high increment of API level by 58.3%. The RMCO implemented continued to increase the API level recorded by 15.8% until August 31st. The total decrement of API level was recorded in Seri Manjung, Kangar and Kuantan (suburban areas) during MCO I, MCO II and MCO III, however, only in between 46.5 – 48.6%.

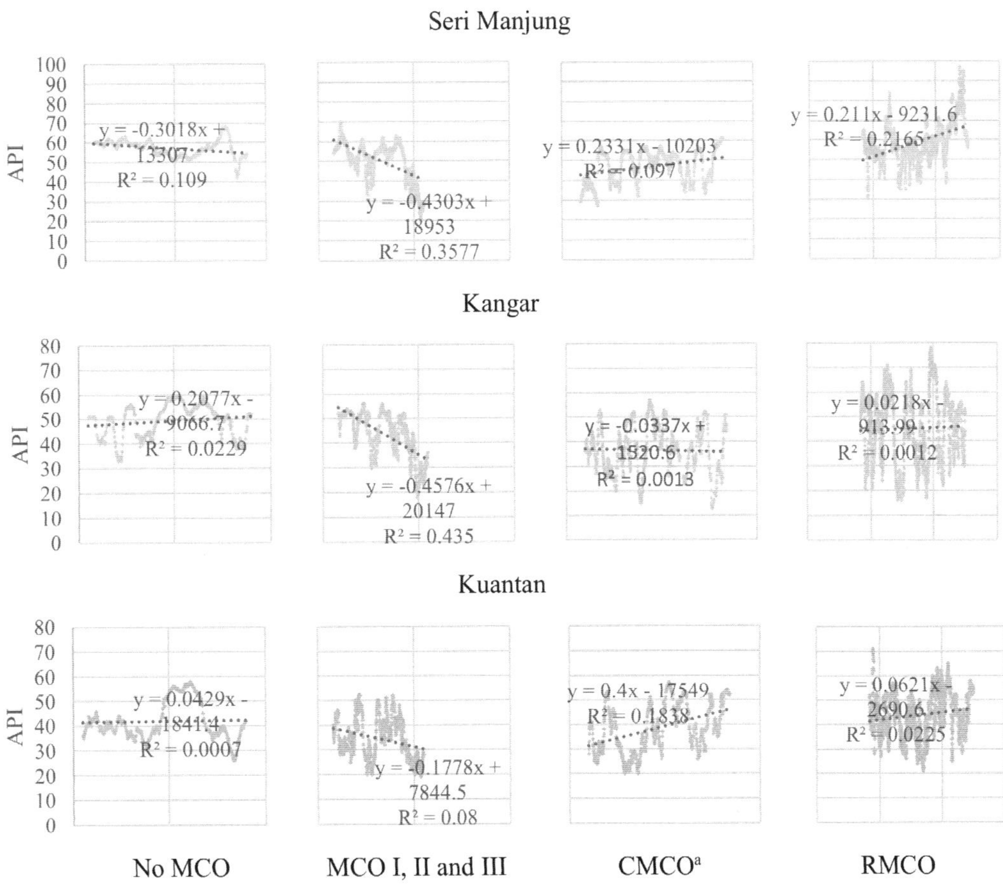

[a] except for Kuantan under MCO IV

Figure 5. Linear regression of API for suburban areas; Seri Manjung (top), Kangar (middle) and Kuantan (bottom) before MCO, during MCO and after MCO (CMCO and RMCO)

The mitigation measures established to curb the transmission of COVID-19 has resulted in the improvement of air quality worldwide. A study on air quality of industrial cities conducted by Nigam et al. (2021) in western India has revealed different patterns of gradual to rapid reduction in most of the pollutant concentrations including PM_{10}, $PM_{2.5}$, carbon monoxide (CO) and sulphur dioxide (SO_2), and later increased as the restrictions were eased. Another study performed by Mandal & Pal (2020) in Delhi, India, has claimed that the air pollution levels dropped instantly after forced restrictions on outdoor activities. Compared to previous years, the study that dealt with six criteria air pollutants including particulate matter, nitrogen dioxide (NO_2), ozone, sulphur oxide and CO from the year 2017 to 2020 shown decrements in the concentrations of the selected pollutants during the lockdown period.

In China, the outcomes of the study conducted by Wang & Su (2020) shown the reduction in carbon emission and a short time improvement in air quality. The country also recorded a reduction in the concentration of NO_2 in the atmosphere. Meanwhile, Dantas et al. (2020) has compared the concentrations of major pollutants during pre-lockdowns and lockdown conditions in Rio de Janeiro, Brazil. From the study, CO that relate to light vehicular emissions shown the most significant reduction together with NO_2 and particulate matter. In a different study considering the pre-virus outbreak and lockdown period, Son et al. (2020) revealed that the concentrations of $PM_{2.5}$ decreased in seven out of ten US states and in the District of Columbia. These studies have shown the major decrement of emissions was related to traffic-induced air pollution [20].

Table 2. Air Quality Status based on API at urban and suburban areas

	No MCO	MCO I	MCO II	MCO III	CMCO[a]	RMCO
Urban area						
Bandaraya Melaka	Moderate	Good	Good	Good	Moderate	Moderate
Cheras	Moderate	Moderate	Moderate	Good	Moderate	Moderate
Alor Setar	Moderate	Moderate	Good	Good	Moderate	Moderate
Suburban area						
Seri Manjung	Moderate	Moderate	Moderate	Good	Moderate	Moderate
Kangar	Moderate	Moderate	Good	Good	Good	Moderate
Kuantan	Good	Good	Good	Good	Good	Good

[a] except Alor Setar and Kuantan under MCO IV

The implementation of MCO has indeed improved the air quality in both urban and suburban areas in Malaysia, especially after further extension of lockdown days. All six study locations have recorded 'good' air quality status during MCO III, which imply low pollution without any bad effect on human health. Bandaraya Melaka (urban area) recorded 'good' air quality status fromthe beginning of MCO I until MCO III. Cheras (urban area), a city located in Kuala Lumpur only recorded 'good' air quality status after the third extension of lockdown, MCO III. Compared to Bandaraya Melaka and Alor Setar, Cheras which is located in the Federal Territory of Kuala Lumpur has a higher population density as well as a large economic growth. According to Akbostanci et al., (2009), the growth of the economic scale results in a proportional growth in environmental pollution. Due to the continuous economic activities before lockdown, the decrement of air quality in this area might take some time after the implementation of a few phases of MCOs. Alor Setar (urban area), located in the northern part of Peninsular Malaysia improved to 'good' air quality status during the second extension of lockdown, MCO II.

Like Cheras, Seri Manjung (suburban area) recorded 'good' air quality status only during MCO III. This is due to the businesses and industries in Perak are focused on some areas including Seri Manjung. Meanwhile, Kangar (suburban area) has improved the air quality status and recorded 'good' during MCO II like Alor Setar (urban area). Kuantan has shown 'good' air quality status before the beginning of MCO, during full lockdown days and the recovery phase. Except for Kuantan, both urban and suburban areas have returned to 'moderate', which showing moderate pollution that does not pose any bad effect on human health, once the RMCO was implemented all over the country.

4. Conclusion

Strict regulations implemented during series of MCOs, followed by the new and revised regulations under the enforcement of the CMCO and RMCO has influenced the transition of the API and air quality status in Malaysia. In this study, it is certainly shown that the air quality in both urban and suburban areas has shown some improvements after the further extension of lockdown days. The implementation of lockdown measures not only restricted the spread of infection rate but has reduced ambient air pollutants levels that lead to improved air quality. The air quality trends, however, gradually deteriorate and has increased based on the implemented measures and the present country's pandemic situation from time to time. As observed by previous studies worldwide, the reduction in pollutant levels related to the lockdown measures has improved the air quality in most parts of the Earth. The findings of this study undoubtedly can offer a potential scope in planning air pollution reduction strategies as Malaysia is making plans to shift to a new stage, living with COVID-19. The COVID-19 outbreak is undeniably disruptive to society, but it has also provided unparalleled opportunities for future scientific development and actions, particularly related to environmental impacts.

Acknowledgements

The research was funded by Universiti Teknologi MARA (UiTM) Johor Branch under the *Geran Penyelidikan Bestari Fasa* 2/2020 (GBest Phase 2/2020) research grant scheme (600-UiTMCJ (PJIA. 5/2)). We highly appreciate the noble contributions and sacrifices by frontliners who have been working relentlessly during this COVID-19 pandemic since March 2020.

References

[1] Borneo Post Online 2020 First coronavirus cases in Malaysia: 3 Chinese nationals confirmed infected, quarantined in Sungai Buloh Hospital. Available from: https://www.bernama.com/en/general/news_covid-19.php?id=1811373
[2] Bernama 2020 First case of Malaysian positive for coronavirus. Available from: https://www.bernama.com/en/general/news_covid-19.php?id=1811373
[3] John B 2020 PM: Malaysia under movement control order from Wed until March 31, all shops closed except for essential services. Malay Mail [Internet]. Available from: https://www.malaymail.com/news/malaysia/2020/03/16/pm-malaysia-in-lockdown-from-wed-until-march-31-all-shops-closed-except-for/1847204
[4] Sukumaran T 2020 Coronavirus: Malaysia in partial lockdown from March 18 to limit outbreak. South China Morning Post [Internet]. Available from: https://www.scmp.com/week-asia/health-environment/article/3075456/coronavirus-malaysias-prime-minister-muhyiddin-yassin
[5] Malaysian National Security Council (NSC) 2020 Arkib SOP 2020 [Internet]. 2021 [cited 2020 Aug 26].Available from: https://www.mkn.gov.my/web/ms/arkib-sop-2020/
[6] Ho Y 2020 Malaysia Braces for Coronavirus Infections to Peak in Mid-April. Bloomberg [Internet]. 2020; Available from: https://www.bloomberg.com/news/articles/2020-04-03/malaysia-braces-for-coronavirus-infections-to-peak-in-mid-april

[7] Bernama 2020 MCO extended another two weeks to 12 May - Muhyiddin
[8] Bernama 2020 Perintah Kawalan Pergerakan bersyarat akan dikenakan - Muhyiddin. Available from: https://www.bernama.com/bm/news.php?id=1837419
[9] Loo C 2020 CMCO ends June 9, Recovery MCO from June 10 to Aug 31 (Updated). The Sun Daily
[10] Yusof A 2020 Malaysia's recovery movement control order extended to Dec 31, tourists still not allowed in: PM Muhyiddin. Channel News Asia
[11] Latif MT, Dominick D, Ahamad F, Khan MF, Juneng L, Hamzah FM and Nadzir MSM 2014 *Sci Total Environ.* **482-483** 336-348
[12] Abdullah S, Ismail M, Ahmed AN and Abdullah AM 2019 *Atmosphere* **10** 667
[13] Abdullah S, Mansor AA, Napi NNLM, Mansor WNW, Ahmed AN, Ismail M and Ramly ZTAR 2020 *Sci Total Environ.* **729** 130922
[14] Department of Environment Malaysia (DoE). Air Pollutant Index (API) [Internet]. 2021. Available from: https://www.doe.gov.my/portalv1/en/info-umum/english-air-pollutant-index-api/100
[15] Nigam R, Pandya K, Luis AJ, Sengupta R and Kotha M 2021 *Sci Rep.* **11** 4285
[16] Mandal I and Pal S 2020 *Sci Total Environ.* **732** 139281
[17] Wang Q and Su M 2020 *Sci Total Environ.* **728** 138915
[18] Dantas G, Siciliano B, França BB, da Silva CM and Arbilla G 2020 *Sci Total Environ.* **729** 139085
[19] Son JY, Fong KC, Heo S, Kim H, Lim CC and Bell ML 2020 *Sci Total Environ.* **744** 141012
[20] Machado S and Mehnaz S 2020 *Environ Sustain.* 1-3
[21] Akbostanci E, Türüt-Aşik S and Tunç GI 2009 The relationship between income and environment in Turkey: Is there an environmental Kuznets curve? *Energy Policy* **3** 861-867

IOP Conf. Series: Earth and Environmental Science 1013 (2022) 012008 doi:10.1088/1755-1315/1013/1/012008

Exposure to industrial air pollution and its association with respiratory symptoms among children in Parit Raja, Batu Pahat

Aida Farihah Mohmad Shamsuddin[1], Juliana Jalaludin[1*] and Nor Haslina Hashim[2]
[1]Department of Environmental and Occupational Health, Universiti Putra Malaysia (UPM), Malaysia
[2] Occupational Safety, Health and Environment Office, Universiti Tun Hussein Onn (UTHM), Malaysia
*Corresponding author: juliana@upm.edu.my

Abstract. Outdoor air pollution for anthropogenic sources is generally caused by the emission of the combustion process which are industries, motor vehicles, and others. These industrial areas emitted a lot of hazardous gases into the atmosphere, polluting the surrounding air to its worse. The close proximity from the industry and duration of stay at the current area could worsen the respiratory health of the children. Compared to adults, children were more likely to get worse implications when exposed to air pollution. This cross-sectional comparative study aimed to determine the exposure to industrial air pollution its association with respiratory symptoms among children aged 3 to 12 years old in Parit Raja, Batu Pahat. Children who fulfilled the inclusion criteria were selected to participate in this study. Those who reside less than 5 km from the industrial area were chosen as the exposed group, while children living more than 5 km from the industrial areas were selected as the comparative group. A questionnaire adapted from the American Thoracic Society was used to determine children's background information and respiratory health symptoms. Nature of area, distance from the main road and industrial area, type of house, building materials, and carpet usage were found significant among the groups. There was a significant difference in respiratory health symptoms of cough and phlegm for the exposed and the comparative group at $p = 0.038$ and $p = 0.008$, respectively. Phlegm was revealed to be associated with the duration of stay in the current area for the exposed group (OR = 5.149, 95% CI = 1.080 – 18.365). On the other hand, phlegm and cough had an association with distance from the industrial area at (OR = 3.889, 95% CI = 1.004 - 15.068) and (OR = 6.707, 95% CI = 1.406 - 31.989), respectively. In multiple logistic regression done on phlegm, it was found that duration of stay (years) was the most significant variable after controlling all confounders at (B = 3.469, $p = 0.004$, AOR = 3.094, 95% CI = 3.108 – 31.353).

1. Introduction

Batu Pahat is located on the northwest coast of Johor, Malaysia, one of which consists of several towns, one of which is Parit Raja. It is a small town but consists of an educational hub (Universiti Tun Hussein Onn Malaysia), industrial, residential and commercial areas. Besides, Parit Raja Industrial Area comprises various manufacturing businesses including wood, electric, textile, plastic and paper container, adhesive and resin manufacturing. These industrial areas are one of the sources of air pollutions that coming from man-made activities. As a result, emitted a lot of hazardous gases into the atmosphere and polluting the surrounding air to its worse.

Due to the increasing population coming from the educational hub and industrial area, Parit Raja becomes a site of urbanisation and has lots of residential areas being built around the industrial zone [15].The growing of transportation and industrial sector around Parit Raja may contribute to the high level of pollutant, in which it is dangerous to health. The proximity between the industrial area with populated regions, making it difficult to avoid the population from being exposed and getting health implications from the pollution. Physical observation also has shown that one industry has the frequent production of plume that may contribute to the high level of PM_{10} [15]. Consequently, study have proven that increased concentration of air pollution will increased the result of Consequently, studies have proven that increased air pollution concentration will increase the result of urbanisation and vehicles caused by the Malaysian population [15].

Figure 1: *The industrial zone is in the center of the city, flanked by the university and residential regions. (Image Generated* by Google maps, 2021)

The exposure to air pollution could first cause acute health effects, which later could develop to a more chronic health effect [5]. Most studies have also mentioned the negative impact of air pollution on children's lung function and respiratory symptoms [7] [12]. In addition, children are especially vulnerable and at risk of getting respiratory health problems because their lungs are not fully developed yet [4]. Therefore, compared to adults, they were more likely to get worse implications when exposed to air pollution since they breath more rapidly, thus absorbed more pollutants [16].

Also, it is such a concern that children who reside near an industrial area and longer in time are continuously being exposed to the air pollutant, because the pollutant can infiltrate into homes and buildings to a greater degree. Therefore, this study aims to determine the exposure to industrial air pollution in terms of duration of stay in the area and distance from the industrial area and its association with respiratory symptoms among children in Parit Raja, Batu Pahat.

2. Methods

2.1 Study Background

This was a cross-sectional comparative study conducted among 104, male and female children aged 3 to 12 years old. The respiratory symptoms of cough, phlegm, wheezing, and chest tightness was used to know the effect of being exposed to industrial air pollution.

2.2 Sampling Technique

Sample size was calculated using the formula on the anticipated outcome in proportions of events in two groups [11], with a total sample size of 104 respondents after increasing to 20% of respondents due to the non-response rate. Children who fulfil the inclusion criteria set by the researchers will be recruited through purposive sampling. Those who live in the study area for less than one year, did not get approval from parents, or did not fill up the questionnaire completely were excluded from participating in this study.

The residential areas were chosen based on the distance from point exposure, which believed to contribute to industrial air pollution. Based on purposive sampling, children who reside within a 5 km radius from the industrial zone were selected as the exposed group. Conversely, children who reside in a residential area with more than a 5 km radius from industrial areas were chosen as a comparative group.

2.3 Questionnaire Form

In addition, a set of adapted questionnaires from the American Thoracic Society (ATS) for children was used and translated in Malay. It consists of 6 sections: i) Background Information of Children and Family Member; ii) House Condition; iii) Outdoor Environment Condition; iv) Children's Health Condition; v) Other Diseases or Current Symptoms; vi) Family Health History. Parents or guardians were asked to complete the questionnaire on behalf of their children. The questionnaires were distributed through physical and online distribution, to reduce physical contact during the COVID-19 situation in Malaysia.

2.4 Exposure Assessment

Moreover, secondary data on the wind rose diagrams indicating the pollution concentration dispersed towards UTHM and the residential areas was used to identify which areas were exposed to industrial air pollution (unpublished data). The wind rose diagram and pollution rose shows a low wind speed cause a higher probability of air pollutants dispersed downwind that is toward the east and southeast from the source of air emissions. The air modelling results also show that community in UTHM and the residential area within 5 km were exposed to high pollutant concentration. It was then compared to follow the residential areas to find the most accurate area to be selected.

2.5 Quality Control

The questionnaire was validated and adapted from a previous researcher. It was also pretested on at least 10% of the respondents to assess the understanding of the respondents on the questionnaires given. It was prepared in the Malay Language for quick and easy understanding. Therefore, the researcher had fully understood the questionnaire requirements and details before the data were collected.

2.6 Statistical Analysis

Data collected in this study were analyzed by using Statistical Package Social Science IBM Version 26 (SPSS Ver. 26) and summarised in the descriptive statistics. Comparison between background information, house condition, the prevalence of respiratory symptoms, association between duration of stay (year) and respiratory symptoms, and the association between distance from residential areas was tested with Mann Whitney U test and Chi-Square test and test with Yates 'correction for an expected value less than 5. In addition, Logistic Regression was used to determine the main variables that influence respiratory health symptoms among study respondents after controlling all the confounders. The p-value of less than 0.05 was considered statistically significant.

2.7 Ethical Consideration

The approval was obtained from the Ethics Committee for Research Involving Human Subjects Universiti Putra Malaysia. Permission was obtained from the responsible party in order to distribute the questionnaire. Also, an informed consent form was distributed to the parents or guardians of the participants, either hardcopy or softcopy. A brief explanation was given to them before filling up the questionnaire. To engage in this evaluation, respondents must be voluntarily oriented. Most crucially, the participants' identity was treated with confidentiality.

3. Results

3.1 Response rate

During the questionnaire distribution, it was distributed through two methods, online and physical. 108 questionnaires were distributed through the message of URL, and 200 questionnaires and consent forms were distributed physically for both exposed and comparative groups. From 108 questionnaires, 50 (46.2%) were approved and responded by parents while the other 58 (53.7%) were excluded as they did not meet the inclusion criteria. Of the 50 respondents, 21 (42%) were from the exposed area, while other 29 (58%) were from the comparative area. For the physical distribution, only 54 questionnaires were returned, which was about 27%. Of the 54 respondents, 31 (57.4%) were from the exposed group, while 23 (42.6%) from the comparative group. Thus, a total of 104 respondents participated in this study. The total response rate for questionnaire distribution was 73.2%.

3.2 Demographic and background information of respondent's residence

Table 1 shows the gender distribution of males in the exposed group were 31 (59.6%) while in the comparative group were 29 (55.8%). The female respondents were relatively higher in the comparative group with 23 (44.2%) and 21 (40.4%) in the exposed group. However, all of the respondents are Malay for both groups. For the age of the exposed and comparative group, the median age was the same, which were 6 for both groups but with different IQR which were 4 for the exposed group and 5 for the comparative group. No significant difference was found in gender, age and race.

Most of the respondents in the exposed group live in a suburban area with 31 (59.6%). On the other hand, most respondents from the comparative group live in a rural area 37 (71.2%). A chi-square test was done on the nature of the area with respondent groups. It shows a significant association between respondent groups with the nature of the area ($p<0.001$). It also shows the duration of stay in the current area for both groups. The median and interquartile range of stay duration for the exposed and comparison groups were 6.00 (2.75) and 6.00 (6.00), respectively. Mann-Whitney U result shows that there was no significant difference in duration of stay and age for both groups.

For a distance of the house from the main road, most residents from both groups reside more than 100 m to 500 m, around 22 (42.3%) for exposed and 20 (38.5%) for comparative groups. However, there was a significant association between distance from the main road ($p=0.024$) and respondent groups. For distance from industrial areas, the exposed group was determined from a distance of less than 5 km from the industrial area. 42 (80.8%) of respondents live within 2.1 km to 5 km and the other 10 (19.2%) live less than 1 km from the industrial area. Conversely, the comparative group was determined from the distance of more than 5 km from the industrial area, with all 52 (100.0%) respondents living more than 5 km from the industrial area. However, there was a significant association between distance from the industrial area ($p=0.000$) and respondent groups.

For the type of house, 22 (42.3%) respondents from the exposed group live in a one-storey terrace house. Respondents from the comparative group mostly live at the village house, with around 28 (53.8%). There was a significant association between type of house at $p < 0.001$ for both groups. The analysis shows that 46 (88.5%) and 35 (67.3%) respondents from respective groups have used carpets at their residences. There was a significant association on carpet usage in both groups.

Table 1. Demographic information and residentials' background information

	N (%) / Median (IQR)		
	Exposed	**Comparative**	**p-value**
Gender			
Male	31 (59.6)	29 (55.8)	0.691
Female	21 (40.4)	23 (44.2)	
Race			
Malay	52 (100.0)	52 (100.0)	-
Age (year)	6.00 (4.00)	6.00 (5.00)	0.842
Nature of Area			
Urban	4 (7.7)	0 (0)	
Sub-urban	31 (59.6)	15 (28.8)	**<0.001****
Rural	17 (32.7)	37 (71.2)	
Duration of stay (year)	6.00 (2.75)	6.00 (6.00)	0.626
Distance from Main Road			
Less than 100 m	20 (38.5)	9 (17.3)	
100 m to 500 m	22 (42.3)	20 (38.5)	**0.024***
501 m to 1000 m	7 (13.5)	15 (28.8)	
More than 1000 m	3 (5.8)	8 (15.4)	
Distance from Industrial Area			
Less than 1 km	10 (19.2)	0 (0)	
1 km to 5 km	42 (80.8)	0 (0)	**<0.001****
More than 5 km	0 (0)	52 (100.0)	
Type of house			
Two-storey terrace house	12 (23.1)	4 (7.7)	
One-storey terrace house	22 (42.3)	15 (28.8)	
Village house	16 (30.8)	28 (53.8)	**<0.001****
Bungalow	0 (0)	5 (9.6)	
Others	2 (3.8)	0 (0)	
Carpet Usage			
Yes	46 (88.5)	35 (67.3)	**0.009***
No	6 (11.5)	17 (32.7)	

*Significant at p<0.05; ** Significant at p<0.001

3.3 Comparisons of respiratory health symptoms among groups

Table 2 shows the comparison of reported respiratory symptoms among children in both groups. 10 (16.2%) of respondents from the exposed group while only 3 (5.8%) from the comparative group reported having cough. In addition, 11 (21.2%) and 2 (3.8%) of respondents from the respective groups were reported by parents to have phlegm symptoms. Prevalence of wheezing was found among 3 respondents (5.8%) of the exposed group and 4 respondents (7.7%) from the comparative group. Meanwhile, the prevalence of chest tightness was found among 6 (11.5%) and 2 (3.8%) respondents from both exposed and comparative groups. The result from the statistical analysis shows that there was a significant difference for cough (x^2 = 4.308, p = 0.038) and phlegm (x^2 = 7.121, p = 0.008) for both groups.

Table 2. Comparison of reported respiratory symptoms among children in both groups

Respiratory Symptoms	N (%)		x^2	p-value
	Exposed Group	Comparative Group		
Cough				
Yes	10 (16.2)	3 (5.8)	4.308	**0.038***
No	42 (80.8)	49 (94.2)		
Phlegm				
Yes	11 (21.2)	2 (3.8)	7.121	**0.008***
No	41 (78.8)	50 (96.2)		
Wheezing §				
Yes	3 (5.8)	4 (7.7)	0.000	1.000
No	49 (94.2)	48 (92.3)		
Chest Tightness §				
Yes	6 (11.5)	2 (3.8)	1.219	0.270
No	46 (88.5)	50 (96.20)		

N = 104; § By x^2 test with Yates 'correction for expected value < 5; *Significant at p<0.05

3.4 Association between duration of stay in the current area and respiratory health symptoms

Table 3 below shows the duration of stay in the current area and respiratory symptoms according to group, either exposed or comparative group. A median was used to categorize the duration of stay in the area, with a median of 6 years. Thus, it was categorised into those who stay less than 6 years and 6 years and above. Result from Table 2 of the Chi-Square Test and test with Yates 'correction for expected value < 5 shows that only phlegm in the exposed group was significantly associated and shows an increased risk with duration of stay at (p = 0.025, OR = 5.15, 95% CI = 1.08 – 18.37). However, neither significant association of cough, wheezing and chest tightness was found in both exposed and comparative groups.

Table 3. Association between duration of stay and prevalence of respiratory symptoms among groups

Variables	Location		N (100%)		p-value	OR	95% CI
			<6 years	≥6 years			
Cough	Exposed	Yes	5 (15.6)	2 (25.0)	0.404	0.56	0.14 – 2.23
		No	27 (84.4)	15 (75.0)			
	Comparative§	Yes	2 (7.7)	1 (3.8)	1.000	2.09	0.18 – 24.51
		No	24 (92.3)	25 (96.2)			
Phlegm	Exposed	Yes	8 (28.1)	2 (10.0)	**0.025***	5.15	1.08 – 18.37
		No	23 (71.9)	18 (90.0)			
	Comparative§	Yes	2 (7.7)	0 (0.0)	0.471	-	-
		No	24 (92.3)	26 (100.0)			
Wheezing	Exposed§	Yes	2 (6.3)	1 (5.0)	1.000	1.27	0.11 – 14.95
		No	30 (93.8)	19 (95.0)			
	Comparative§	Yes	3 (11.5)	1 (3.8)	0.603		0.32 - 33.61
		No	23 (88.5)	25 (96.2)			
Chest Tightness	Exposed§	Yes	4 (12.5)	2 (10.0)	1.000	1.29	0.21 – 7.76
		No	28 (87.5)	18 (90.0)			
	Comparative§	Yes	2 (7.7)	0 (0.00)	0.471	-	-
		No	24 (92.3)	26 (100.0)			

N = 104; § By x^2 test with Yates' correction for expected value < 5; *Significant at p<0.05

3.5 Association between distance from the residential area to the industrial area and respiratory health symptoms

Table 4 below shows the relationship between distance from a residential area to an industrial area with the respiratory symptoms among children. A statistical analysis shows distance from residential to industrial area have a significant difference in respiratory symptoms of cough and phlegm at (p = 0.038, PR = 3.89, 95% CI = 1.00 - 15.07) and (p = 0.008, PR = 6.71, 95% CI = 1.41 - 31.99) respectively. However, no significant association was found on wheezing and chest tightness in both groups.

Table 4. Association between Distance from Industrial Area and Prevalence of Respiratory Symptoms

Variables	**N (%)**		**p-value**	**OR**	**95% CI**
	<5 km from industry	**≥ 5 km from the industry**			
Cough					
Yes	10 (16.2)	3 (5.8)	**0.038***	3.89	1.00 - 15.07
No	42 (80.8)	49 (94.2)			
Phlegm					
Yes	11 (21.2)	2 (3.8)	**0.008***	6.71	1.41 - 31.99
No	41 (78.8)	50 (96.2)			
Wheezing §					
Yes	3 (5.8)	4 (7.7)	1.000	0.74	0.16 - 3.46
No	49 (94.2)	48 (92.3)			
Chest Tightness §					
Yes	6 (11.5)	2 (3.8)	0.270	3.26	0.63 - 16.97
No	46 (88.5)	50 (96.20)			

N = 104; § By x^2 test with Yates 'correction for expected value < 5; *Significant at p<0.05

3.6 The main variables that influence phlegm among study respondents after controlling all confounders

Logistic regression was conducted to determine the main variables that influence phlegm after controlling all confounders. The confounders included were sociodemographic factors, the background of respondent's residences, house environment, indoor air pollutant source at home and from activity at home, and smoker. Table 5 shows that the duration of stay at the current area was the most significant factor in children that influenced respiratory symptoms of phlegm among children at (B = 3.469, p = 0.004, AOR = 3.094, 95% CI = 3.108 – 31.353). The result shows that duration of stay at the current area influenced 41.3% (Nagelkerke R^2= 0.413) of getting phlegm symptoms.

Table 5. Logistic regression for factors that influence respiratory symptom (phlegm) after controlling all confounders

	B	S.E.	p-value	AOR	95% CI
Constant	0.462	1.816	0.799	1.587	-
Duration of Stay at Current Area	3.469	1.191	**0.004***	3.094	3.108 – 31.353
Distance from Industrial Area	-1.681	1.163	0.148	0.186	0.019 - 1.818
Nature of Area	-1.582	1.622	0.329	0.206	0.009 - 4.937
Use of Mosquito Repellent	-0.640	0.832	0.442	0.527	0.103 – 2.696
Use of Carpet	1.296	0.964	0.179	3.892	0.574 – 26.391
Distance from Main Road	-1.000	1.471	0.497	0.368	0.021 – 6.574

* Significant at p<0.05
95%CI = 95% Confidence Interval, B = Regression Coefficient, S.E. = Standard Error, Nagelkerke R^2= 0.413

4. Discussion

4.1 Demographic and background information of respondent's residence

A total of 104 respondents were recruited in this study. Half of the respondents were responded through hardcopy, while the others were through online approach. The sample size was increased to 20% to lessen the response rate's impact. Children whose' parents or guardians completed the entire questionnaire distributed either online or hardcopy were chosen.

Table 1 shows there was a significant difference in the nature of area between both groups. This made the risk of respiratory symptoms higher among children in the exposed area as there is a high pollutant level from outdoor sources. Moreover, those in the exposed group were most likely from the suburbs, as Parit Raja Industrial Area was within the suburbs area.

There was a significant difference between the distance from the main road among the study groups at (χ^2 = 9.449, p=0.024). A similar study also found that children who live near a big road or a high-traffic region in urban areas have a higher risk of experiencing respiratory symptoms or adverse health effects among the exposed and comparative groups [1] [14].

As for the distance to an industrial area, the exposed group stayed near the industry, contrary to the comparative group. This is because the distance to the industrial area was used to determine the children coming from which group. This result ties well with previous studies wherein there was a significant association between closer proximity on the industrial area between school or residential area from both groups with respiratory symptoms [2] [9].

In addition, there was a significant association between types of the house at $p < 0.001$ for both groups. Around 42.3% of respondents from the exposed group living in a one-storey terrace house, while more than half of respondents from the comparative group live in a traditional village house. The development of terrace house is increased as there are closer to the industrial area. This is because when there is economic growth in a specific location, the more recently built neighbourhoods are within the vicinity. On the other hand, more traditional village dwellings, also known as Malay kampung houses are for individuals who reside more than 5 km from the industrial sector, as there are numerous villages all across Parit Raja, Batu Pahat.

Moreover, there was a significant difference between carpet usage in both groups with x^2 = 6.755 and p = 0.009. The closer proximity to industrial areas, making parents use carpets. The carpet used were collected allergens, dust, and other contaminants in carpets and remained there until they could be fully disposed. Also, they used carpet to give a positive house appearance and enhance the comfort of a room as a place to sit on the floor, especially for children. A similar result shows there was a significant difference in using carpet at home between urban and semi-urban areas, at $p < 0.05$ with 439 respondents (34.0%) [19].

To compare of the duration of stay in the current area for both groups, there was no significant difference in the duration of stay between the two groups. In line with previous studies, the duration of residence at a specific location has no association with groups [1].

4.2 Comparisons of respiratory health symptoms among groups

Children in the exposed group reported to have more reported symptoms of cough, phlegm and chest tightness. There was a significant association found between cough at x^2 = 4.308 and p = 0.038 and phlegm x^2 = 7.121 and p = 0.008 between exposed and comparative groups. Nevertheless, no association was found for wheezing and chest tightness in both groups. Children that live closer to the industrial area were most likely to have been exposed to a greater level of air pollutants, resulting in an increased risk of getting respiratory symptoms [2] [3]. A residential area close to the main road can increase the risk of getting exposed to pollutants compared to those living far from the main road.

A study by [10] on school children aged 11 years old revealed a significant relationship in reported respiratory symptoms in both groups. A higher prevalence of cough was found among 57 respondents from exposed and 31 respondents from comparative. Likewise, a higher prevalence of phlegm was found among 31 respondents from exposed and 13 respondents from comparative. Another study by [6] also proven that children who home or school less than 2 km from chipboard industries had worsened respiratory health outcomes of cough and phlegm than those who live far away.

4.3 Association between duration of stay at the current area and respiratory health symptoms.

As for the association between duration of stay at the current area and respiratory symptoms among respondents in respective groups, a median value (6.00) was calculated and used to categorize the data into < 6 years and ≥ 6 years. As for the result, only phlegm for the exposed group is revealed to have a significant association with duration of stay at the current area with ($p = 0.025$, OR = 5.15, 95% CI = 1.08 – 18.37). Moreover, those who reside longer in their current address may have more prolonged exposure to industrial air pollution, worsening the health implications. Other researchers in their research have revealed that an increase in residential years (≤ 10 years) in children living near the industrial area may have a higher association of phlegm (OR = 1.01, 95% CI = 0.65 – 1.22) and wheezing (OR = 1.01, 95% CI = 1.01 – 1.67) for children residing near the industrial area as compared to the residential area [13].

Exposure to air pollution can exacerbate the lower respiratory infections that include phlegm symptoms. The prevalence of phlegm increases could be the result of increased airway interaction with allergens and pathogens, as well as a decrease in lung defence [17]. Since the lower respiratory symptoms in children are closely related with infections, it is possible that the pathogen's penetration into the lung cells could increase the inflammation, thus resulting in the lower respiratory infections. [17].

4.4 Association between distance from the residential area to the industrial area and respiratory health symptoms.

Since there was limited data on the air pollutants at the ambient level and no current data collected on the industrial emissions, a cut-off point of distance for less than 5 km from industries and more than 5 km from industries was obtained and referred from a previous study [10], who did the study on school and home exposure to the industrial area in Kemaman, Terengganu. It is also based on the wind rose study and air monitoring in UTHM to know the dispersed pollution that originates from the industrial area. From the result in Table 4, statistical analysis shows distance from industrial area have a significant difference in respiratory symptoms of cough and phlegm at ($p = 0.038$, PR = 3.89, 95% CI = 1.00-15.07) and ($p = 0.008$, PR = 6.71, 95% CI = 1.41-31.99) respectively.

Since cough and phlegm show a significant association between the exposed and comparative group, it clearly shows that cough and phlegm were usually the symptoms of upper or lower respiratory infection. A study discussed that coughing is an action of body defence mechanism in which the body react to remove the phlegm from the lungs and bronchi when pollutant enter the respiratory tract [8]. Phlegm that being produced indicates there was an increased inflammatory response due to contaminant's exposure [8]. As a result, the surge in symptoms of cough and sputum implies that children in industrial regions are more exposed to air pollution than those in the comparative area [8].

In addition, a study found the significant a significant relationship and higher prevalence of cough and phlegm found in those who live less than 5 km from the industrial area [10]. In contrast, another study chose the children living 1 km from petrochemical industries as a studied group while those living far, which is more than 10 km from petrochemical industries, selected as a comparative group [2]. It was found that there were a higher prevalence and significant association found between respiratory health outcomes for those who were living 1 km from the petrochemical industry have 5 times more likely to get cough as compared to those who live far (PR = 5.09, $p = 0.001$, 95% CI = 2.23 – 11.65). The exposed respondents were 9 times as likely to experienced phlegm as the comparative group (PR = 9.66, $p = 0.001$, 95% CI = 2.10 – 44.6) [2].

4.5 The main variables that influence phlegm among study respondents after controlling all confounders.

Logistic regression shows that the most significant factor associated with phlegm symptoms was the duration of stay in the current area. The increased duration of stay at the current area is associated with the increased risk of getting phlegm (AOR = 3.094). The result shows that 41.3% (Nagelkerke R^2= 0.413) of getting phlegm symptoms was influenced by duration of stay at the current area.

Furthermore, those who have lived at their current address for a more extended period of time may be exposed to industrial air pollution for a prolonged period, worsening the health consequences. had Some studies revealed that the increase in length of residence was a significant risk factor for respiratory symptoms [18]. There was also an increased risk of respiratory symptoms among children living in the industrial region as compared with other areas. Additionally, the other study shows the long-term exposure to industrial air pollution have been linked with how many years the family have lived at the current address in the past 5 years [3].

5. Conclusion

In summary, this study found that the higher prevalence of respiratory symptoms among children in exposed area, resulting from the exposure to industrial air pollution. Furthermore, the duration of stays in the area have a significant association with phlegm in the exposed group. Meanwhile, distance from the residential area to the industrial area was significantly associated with the risk of getting cough and phlegm symptoms among children living near with the industrial area. Besides, it can be found that duration of stay in the current area is the most significant factor in increasing the prevalence of respiratory symptoms of phlegm among children after controlling all confounders factors.

As for study limitation, since this study did not undertake any outdoor and indoor air pollution measurements at residential or school sites, outdoor and indoor air pollution measurements at residential or school sites, or any lung function test, the current exposure may not be established. Therefore, the association can only be done based on the responses to the questionnaire and the secondary data available. Furthermore, the parameter was not measured because of the Covid-19 situation in Malaysia, which is worsening by the day, and the Movement Control Order (MCO) established at Parit Raja during the data collection period as a preventive precaution.

As for recommendations, this suggests that regulatory authorities do an annual awareness program that emphasizes air pollution and its health consequences when exposed to it. Industries should also comply with the laws and regulations of the government and come out with more effective control measures to reduce the pollutant dispersed towards residential area nearby. In addition, parents should ensure the cleanliness of their house by cleaning regularly and should consider installing air conditioners or air purifier in residences to limit pollutant dispersion from the outdoors. Finally, to improve this research in the future, consider monitoring air pollution monitoring for factories and indoor exposure to achieve a stronger association between industrial air pollution exposure and children's respiratory health.

Acknowledgement

The authors would like to thank all respondents who volunteered to participate in this study and their cooperation throughout the data collection process.

References

[1] Asa, P., & Jinsart, W. (2016). Effects of air pollution related respiratory symptoms in schoolchildren in industrial areas Rayong, Thailand. *EnvironmentAsia*, *9* (1), 116–123.

[2] Ayuni, Juliana, & Ibrahim. (2014). Exposure to PM 10 and NO 2 and Association with Respiratory Health among Primary School Children Living Near Petrochemical Industry Area at Kertih, Terengganu. *Journal of Medical and Bioengineering*, *3*(4).

[3] Bergstra, A. D., Brunekreef, B., & Burdorf, A. (2018). The effect of industry-related air pollution on lung function and respiratory symptoms in school children. *Environmental Health: A Global Access Science Source*, *17*(1).

[4] Buka, I., Koranteng, S., & Osornio-Vargas, A. R. (2006). The effects of air pollution on the health of children. *Paediatrics and Child Health*, *11*(8), 513–516.

[5] Chen, Z., Salam, M. T., Eckel, S. P., Breton, C. V., & Gilliland, F. D. (2015). Chronic effects of air pollution on respiratory health in Southern California children: Findings from the Southern California Children's Health Study. *Journal of Thoracic Disease*, *7*(1), 46–58.

[6] de Marco, R., Marcon, A., Rava, M., Cazzoletti, L., Pironi, V., Silocchi, C., & Ricci, P. (2010). Proximity to chipboard industries increases the risk of respiratory and irritation symptoms in children. The Viadana study. *Science of the Total Environment*, *408*(3), 511–517.

[7] Dunea, D., Iordache, S., & Pohoata, A. (2016). *Fine Particulate Matter in Urban Environments : A Trigger of Respiratory Symptoms in Sensitive Children*.

[8] Eom, S. Y., Choi, J., Bae, S., Lim, J. A., Kim, G. B., Yu, S. Do, Kim, Y., Lim, H. S., Son, B. S., Paek, D., Kim, Y. D., Kim, H., Ha, M., & Kwon, H. J. (2018). Health effects of environmental pollution in population living near industrial complex areas in Korea. *Environmental Health and Toxicology*, *33*(1), 1–8.

[9] Kamaruddin, A. S., Jalaludin, J., & Chua, P. C. (2015). Indoor Air Quality and Its Association with Respiratory Health among Malay Preschool Children in Shah Alam and Hulu Langat, Selangor. *Advanced in Environmental Biology*, *9*(9), 17–26.

[10] Kamaruddin, A. S., Jalaludin, J., & Hamedon, T. (2016). Exposure to Industrial Air Pollutants and Respiratory Health School and Home Exposure among Primary School Children in Kemaman, Terengganu. *International Journal of Applied Chemistry*, *12*(1), 45–50.

[11] Lemeshow, S., Jr Hosmer, D. W., Klar, J., & K. Lwanga, S. (1990). *Adequacy of Sample Size in Health Studies*. John Wiley & Sons Ltd.

[12] Li, S., Williams, G., Jalaludin, B., & Baker, P. (2012). Panel studies of air pollution on children's lung function and respiratory symptoms: A literature review. *Journal of Asthma*, *49*(9), 895–910.

[13] Mathew, J., Goyal, R., Taneja, K. K., & Arora, N. (2015). Air pollution and respiratory health of school children in industrial, commercial and residential areas of Delhi. *Air Quality, Atmosphere and Health*, *8*(4), 421–427.

[14] Mohd Nor Rawi, N. A., Jalaludin, J., & Chua, P. C. (2015). Indoor air quality and respiratory health among malay preschool children in Selangor. *BioMed Research International*, *2015*.

[15] Raffee, A. F., Hamid, H. A., Mohamed, R. M. S. R., & Jaffar, M. I. (2018). Time series analysis of PM10 concentration in Parit Raja residential area. *International Journal of Engineering and Technology(UAE)*, *7*(3.23 Special Issue 23), 15–21.

[16] Schwartz, J. (2004). Air Pollution and Children ' s Health. *Pediatrics*, *May 2004*.

[17] Siddique, S., Ray, M. R., & Lahiri, T. (2011). Effects of air pollution on the respiratory health of children: A study in the capital city of India. *Air Quality, Atmosphere and Health*, *4*(2), 95–102.

[18] Wichmann, F. A., Müller, A., Busi, L. E., Cianni, N., Massolo, L., Schlink, U., Porta, A., & Sly, P. D. (2009). Increased asthma and respiratory symptoms in children exposed to petrochemical pollution. *Journal of Allergy and Clinical Immunology*, *123*(3), 632–638.

[19] Zainal Abidin, E., Semple, S., Rasdi, I., Ismail, S. N. S., & Ayres, J. G. (2014). The relationship between air pollution and asthma in Malaysian schoolchildren. *Air Quality, Atmosphere and Health*, *7*(4), 421–432.

Association of short-term exposure to sulphur dioxide and nitrogen dioxide with number of hypertensions in East Java, Indonesia

S A Firdausi[1], R Azizah[1*], J Jalaludin[1,2], Z A Zakaria[1,3]

[1]Department of Environmental Health, Faculty of Public Health, Universitas Airlangga, 60115, Mulyorejo, Surabaya, Indonesia
[2]Department of Environmental & Occupational Health, Faculty of Medicine and Health Sciences, Universiti Putra Malaysia, 43400 Serdang, Selangor, Malaysia
3Department of Biomedical Science, Faculty of Medicine and Health Sciences, Universiti Putra Malaysia, 43400 Serdang, Selangor, Malaysia
*Corresponding author: azizah@fkm.unair.ac.id

Abstract. Previously, several studies conducted related to hypertension which associated with sulphur dioxide and nitrogen dioxide. The number of East Java's hypertension cases from 2015 to 2017 had increased. It is predicted that the number of hypertension cases would be increased. This study was aimed to determine the association between short term exposure to sulphur dioxide and nitrogen dioxide with the number of hypertension cases in East Java. This research was an ecological study, where ambient air pollutants data were taken from the Environmental Management Information Report of East Java 2017, while the number of hypertension cases was taken from the report of Health Profile of East Java Province 2017. Spearman coefficient correlation was used to determine the objective. There was no association between sulphur dioxide with the number of hypertension cases according to gender (male: p= 0.624, r= -0.082; female: p= 0.697, r= -0.065; all cases: p= 0.768, r= -0.050) and nitrogen dioxide (male: p= at 0.24, r= 0.195; female: p= 0.611, r= 0.085; all cases: p= at 0.254, r= 0.190). There was an increase in the number of East Java hypertension cases. However, there was no association between SO_2 and NO_2 with the number of East Java hypertension cases.

Keywords: Outdoor air pollution, sulphur dioxide, nitrogen dioxide, hypertension, East Java Province.

1. Introduction

One of the major causes of health diseases and death in the world is outdoor air pollution [1]. Air pollutants can be originated from natural and anthropogenic sources. There is a potential human contribution to cause greater air pollution compared to natural sources [2]. Air pollution that is related to traffic is a global problem. This is because it can be found in developed, developing and underdeveloped countries [3,4]. Air quality in many large countries has decreased rapidly in line with an increase in population and ownership of motor vehicles [5].

The East Java Province Environmental Status used Indeks Kualitas Udara (IKU) in their annual air quality index. However, the report only measured and reported on short term NO_2 and SO_2 ambient levels. The parameters tested with this method are limited to the NO_2 and SO_2 parameters. This measurement method is a standard method that is currently applied nationally for use in IKU counters in provinces and districts / cities in Indonesia. The scale of the IKU value is in the range of 0-100. The higher the IKU value means the better air quality. IKU in East Java in 2015 amounted to 91.09, down to 83.37 in 2016. In 2017, the IKU increased to 85.49 [6].

The use of energy in daily life can be related to air quality. Most of the industrial and transportation sectors still use fossil fuels such as petroleum and coal 49 [6]. The use of fossil fuels can cause the production of SO_2. Fossil fuels are produced by power plants and other industrial facilities. Other SO_2 sources include industrial activities such as metal production processes, vehicle fuels with high sulphur content and volcanic activity [7–9]. Exhaust gases from burning fossil fuels released gases pollutants such as NO_2 and SO_2 . According to the observing ambient air quality in the district/city, air chemicals such as NO_2 , SO_2 , and carbon monoxide have increased, especially in congested traffic areas such as highways. The biggest contributors to SO_2 , NO_2 , and CO concentrations in the air are motor vehicles [6].

The decline in environmental health is in line with population growth, industry and the number of vehicles on the road. This potentially can reduce health conditions, both acute and chronic [10,11]. Based on East Java Province Health Profile data in Figure 1, the number of hypertension cases from 2015 to 2017 had increased [12–14].

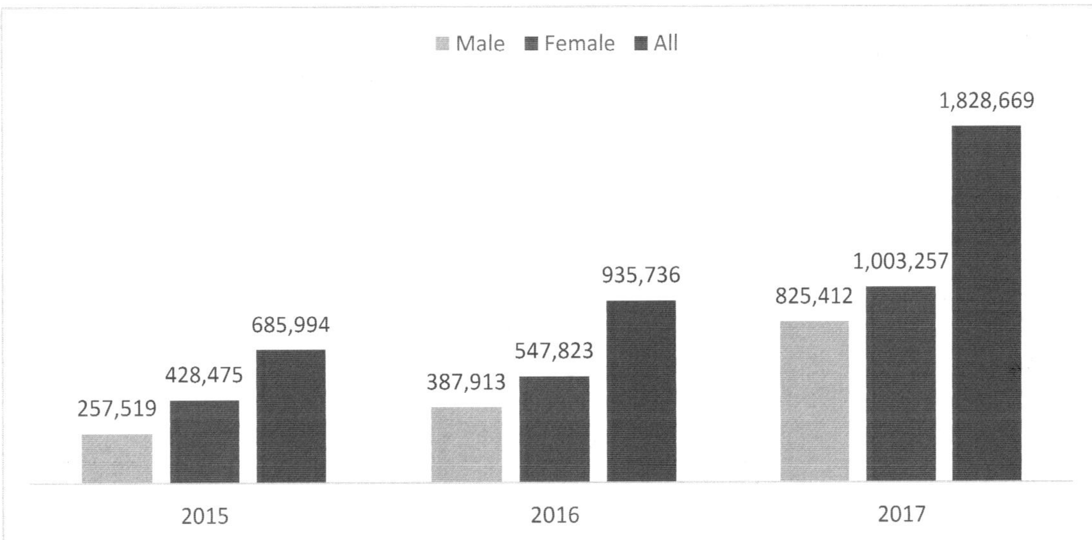

Figure 1. The number of Hypertension Cases in East Java Province from 2015 to 2017.

Based on that, there is a possibility of increase in the number of hypertension cases in East Java to this day. A result of short- or long-term exposures can happen if there are contrary health significances of air pollution [1]. Besides particulates (PM) and Ozone, NO_2 and SO_2 are classified as dangerous pollutants that can affect human health [2].

There were 17 meta-analysis studies conducted. Those were examined the effects of short- and long-term exposure to air pollutants. Short term exposure to SO_2 is considerably related to hypertension, according to the study results [15]. A study conducted in South Korea also showed an association between ambient air pollutants concentrations and blood pressure. The absorption of NO_2 chemicals was significantly associated with the blood pressure in summer, while SO_2 was more likely to be associated with blood pressure in winter [16].

Due to limited information about SO_2 and NO_2 with hypertension in East Java, we conduct research to determine the correlation of short-term exposure to SO_2 and NO_2 with the number of hypertension cases in East Java, Indonesia.

2. Methods

This research used secondary data of ambient air pollution and hypertension cases in 2017 based on Environmental Management Information Report of East Java Province [17], with an ecological study based approach.

Monitoring ambient air quality in East Java Province in 2017 was carried out once a year, in all districts/cities in East Java (38 districts/cities), as many as 2 (two) times a year, with short time duration monitoring, 2 (two) weeks in each time of monitoring. The monitoring program was using the Passive Sampler method, which was a collaboration between the Ministry of Environment and Forestry and the Provincial Environmental Agency and the Regency/City Environmental Agency in East Java. We can see the map of East Java Province in figure 2.

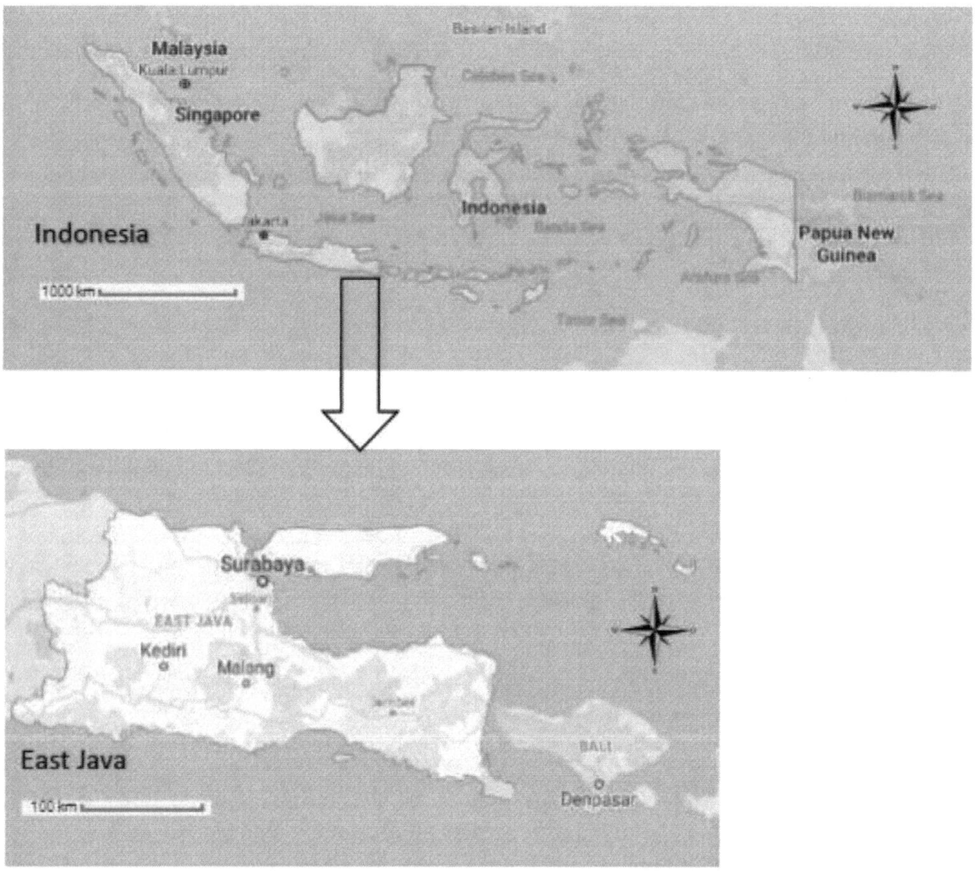

Figure 2. Map of East Java (maps.google.com)

There were two types of pollutants monitored in this study, which were sulphur dioxide (SO_2) and nitrogen dioxide (NO_2). Three monitoring locations in each district/city were selected, representing the residential area, traffic area, and industrial area. Then, the value of pollutant per parameter was summed and the average values were calculated to associate with the number of hypertension cases. This measurement was a standard method which currently applied nationally in calculating the IKU in provinces and districts/cities throughout Indonesia.

The dependent variable was secondary data on the number of adult hypertension cases for one year taken from the Health Profile of East Java Province in 2017 [18]: The number of hypertension cases was grouped into 3 categories: male, female, and all cases. Univariate and bivariate statistical tests were used to determine the association between SO_2 and NO_2 with the number of hypertension cases using

Spearman Coefficient Correlation with SPSS version 25. Based on the profile, p-value < 0.05 was considered significant.

3. Results

Total cases of hypertension in East Java province in 2017 were 1,828,669 cases, where 1,003,257 cases were among female and 825,412 cases were among males. The descriptive statistics for air pollutants concentrations were shown in Table 1.

Table 1 Frequency Distribution of Ambient Air Quality in East Java 2017

Independent Variable	N	Frequency Distribution						EU Standard
		Min.	Max.	Mode	Mean	Median	SD	
SO_2 (ppm)	38	0.00135	0.00551	0.00256	0.0028081	0.0025382	0.00106641	0.00763
NO_2 (ppm)	38	0.00070	0.01247	0.00837[a]	0.0075462	0.0074043	0.00232760	0.02127

The European standard for NO_2 is 40 μg/m³ or 0.02127 ppm and the SO_2 standard is 20 μg/m³ or 0.00763 ppm [19]. As shown in Table 1, the average SO_2 and NO_2 obtained from this study were below the EU standard for both pollutants. In this article the measurement standard is compared to the EU standard, because for IKU measurements in Indonesia, the average numbers NO_2 and SO_2 of the provinces are compared with the EU (European United) reference to obtain the EU air index model (Ieu).

The number of hypertension cases obtained from the Health Profile in one-year monitoring included only patients aged above 18 years old. The frequency distribution of the number of hypertension cases as shown in Table 2.

Table 2 Frequency Distribution of the Number of Hypertension Cases in East Java in 2017

Independent

Independent Variable	N	Frequency Distribution					
		Min. number of cases	Max. number of cases	Mode	Mean	Median	SD
Male	825,412	221	150,756	15,721	21,721.36	15,192.5	28,348.97
Female	1,003,257	721	106,646	11,982[a]	26,401.50	21,832	22,551.79
All cases	1,828,669	1117	231,349	43,879	48,122.86	40,555	47,677.17

Based on Table 2, the male gender has the minimum number of hypertension cases (N=221), compared to the female gender with the number of hypertension cases at N=721. Meanwhile, the maximum number of hypertension cases for males was higher than the female at N= 150,756 and N= 106,646, respectively. However, the average number of female hypertension cases was above the average number of male hypertension cases. The results of the statistical test of association between SO_2 and NO_2 with the number of hypertension cases according to gender were shown in Table 3.

Table 3. Spearman correlation coefficients of air pollutants with the number of hypertension cases.

Ambient Air Pollutant	SO_2		NO_2		Hypertension					
					Male		Female		All	
	P	r	p	r	p	r	p	r	p	r
SO_2	-		0.503	-0.082	0.624	-0.082	0.697	-0.065	0.768	-0.050
NO_2	0.503	0.112	-	-	0.240	0.195	0.611	0.085	0.254	0.190

Based on Table 3, the p-values of SO_2 and NO_2 obtained from the Spearman correlation with the number of hypertension in male, female, and both genders were above 0.05. So, it can be concluded that there was no association between SO_2 and NO_2 with the number of hypertension in East Java in all categories of gender.

4. Discussion

Statistical association test from this study showed that ambient air pollutants were not significantly associated with the number of hypertension cases in male, female and both genders.

The SO_2 in this research was not associated with the number of hypertension in male, female, or all cases of hypertension. However, a study in Taiwan found an association between SO_2 from 1993 to 2005 with cases of hypertension (OR: 1.17, CI 1.10-1.25) (20). This difference is possibly due to the different doses of SO_2 in the air. Research on mice with different SO_2 doses shows differences in the effect of SO_2 or its derivatives on heart function. High concentrations of SO_2 are associated with increased ROS and decreased ATPase activity and the effects of damage to the liver. SO_2 or its derivatives in low concentrations may be able to modulate heart function through the pathway of NO signal transduction. SO_2 gives a higher vasorelaxant effect compared to SO_2 derivatives [21].

The SO_2 is a type of pollutant gas that is commonly found. SO_2 in the air enters the body through the respiratory tract; then, it is easily hydrated to produce sulphuric acid. The next process is dissociation to form its derivatives, namely bisulfite and sulfite (1: 3 M / M, in neutral liquid [22,23]. SO_2 derivatives can enter the entire body through the blood. As for living things, the biological and toxicological effects of exogenous SO_2 are originated from their derivatives. Previous studies showed that the blood pressure of rat decreased due to SO_2 and its derivatives [23,24].

Cardiovascular disease comes from hypertension as an essential risk factor, but its pathogenesis has not been fully explained. Studies in mice with SO_2 exposure (50 ppm, 6 hours/day, 5 days/week for 31 weeks) have shown consistency in blood pressure slowly decreasing insusceptible to salt-induced hypertension rats; it can be interpreted that SO_2 might affect blood pressure [25,26]. Vasorelaxation dysfunction is the major component pathogenesis of hypertension [26]. SO_2 can increase vasorelaxation in the arteries of spontaneous hypertensive rats (SHR) by increasing the vasodilation response to NO in the aortic ring, which increases the production of NO aortic tissue [25, 27]. The interaction between SO_2 and NO is involved in the mechanism by which SO_2 regulates hypertension [25].

NO_2 in this study was also not related to the number of cases of hypertension in East Java in 2017. The results from this study are similar to studies conducted on NO_2 exposure towards coronary heart disease and left systolic ventricular disorders. Their study showed there were no significant changes in blood pressure [28]. The results of this study are similar to those of respondents who were exposed to NO_2 . The results showed that NO_2 inhalation did not damage vascular vasomotor or fibrinolytic function. NO_2 does not appear to be a major cause of adverse cardiovascular effect in air pollution [29].

Researches showed different results between the relationship of NO_2 and the incidence of hypertension. Similar results were previously conducted in the US; the results showed there was no evidence of correlation related to a higher incidence of hypertension with higher NO_2 levels. This finding is possible because of the lower incidence at higher exposure levels [30]. These results differ from studies conducted based on time-series studies in Taiwan, the results showing an association between NO_2 and hypertension cases in a single model (OR: 1.07, CI: 1.00-1.14) [20]. The results of a Canadian study showing a positive correlation of the increase of NO_2 exposure with hypertension (OR = 1.02, 95% CI: 1.01, 1.03) [31].

Measurement of the distribution of air pollution at the city level requires an appropriate model. This method is challenging due to the lack of routine data. There are doubts about the automatic spatial correlation of air pollution and its health impact in the city. This doubt is due to other risk factors that are not measurable, other risk factors that are not represented in the predictions and imperfect estimates of the level of air pollution [32].

This study has some limitations that should be explained. First, the secondary data on ambient air pollution from the Environmental Services report was taken only four weeks within a year and only at three points in each city. So they were less able to represent the air quality level of city/regency in East

Java Province. Other than that, there was no data on time exposure to the pollutants and the hypertension diagnosis. Research with time and personal exposure is required because the effect of air pollutants exposure can vary according to the duration of exposure. The impact of SO_2 exposure can occur rapidly. People who inhale it will acquire severe symptoms within 10-15 minutes [33].

It is important to note, SO_2 emissions are the main precursor of $PM_{2.5}$ ambient. SO_2 can react with other air pollutants and produce sulfate particles. These sulfate particles are constituents of fine particles ($PM_{2.5}$). The effects of $PM_{2.5}$ include cardiovascular and respiratory health [16], so further research is needed on $PM_{2.5}$ and other air pollutants.

Based on the data from this study, we need to increase the implementation of precautions to control and minimize the emissions of SO_2 and NO_2 into the outdoor air for better outdoor air quality and health of its citizens. Reducing air pollution requires cross-sectoral collaboration at the national and regional levels in various fields such as transportation, urban planning, and waste management. Efforts to control sources of pollution correctly need to be done, because if it cannot be done properly, such as improper waste management, it can cause contrary health significances of air pollution [34–36].

Limitation of this study's parameters measured only SO_2 and NO_2. This is due in the East Java Province Environmental Status report the data used in IKU calculations is only SO_2 and NO_2 parameters. So it needs further research that examines other air parameters and examines individual exposure to health impacts. Further studies in this field would be important for a better understanding of the association of SO_2 and NO_2 with hypertension.

5. Conclusion

In summary, based on the data on the ambient air pollutants and the number of hypertension cases obtained from 38 city/residences in East Java for the year 2017, the results showed that there is no correlation of short term exposure to SO_2 and NO_2 with the number of hypertension cases (male: p= 0.624, r= -0.082; female: p= 0.697, r=-0.065; all cases: p= 0.768, r= -0.050) and NO_2 (male: p= 0.24, r= 0.195; female: p= 0.611, r= 0.085; all cases: p= 0.254, r= 0.190), even though the number of hypertension cases has increased for the year 2017 as compared with the previous year. The authors expect that SO_2 and NO_2 emission reduction programs in Indonesia can be increased so that the levels of these substances in ambient air remain below predetermined limits. Further research on the other hypertension risk factors and individual exposure in other areas in Indonesia and its health impacts are needed to give more comprehensive data and results.

Acknowledgements

This research was supported by the International Student Outbound Fulltime Programme conducted by Master of Environmental Health, Faculty of Public Health, Universitas Airlangga in association with Faculty of Medicine and Health Sciences, Universiti Putra Malaysia.

References

[1] WHO 2018 *Ambient air pollution: Health impacts* [Internet] Geneva: WHO; [cited 2019 Nov 12]. Available from: https://www.who.int/airpollution/ambient/health-impacts/en/

[2] WHO 2019 *Ambient Air Pollution: Pollutants* [Internet]. Geneva: WHO [cited 2019 Dec 10]. Available from: https://www.who.int/airpollution/ambient/health-impacts/en/

[3] Whyand T, Hurst JR, Beckles M, Caplin ME 2018 Pollution and respiratory disease : can diet or supplements help ? A review. *Respir Res* [Internet]. [cited 2019 Nov 12];**19**(1)1–14. Available from PubMed: https://pubmed.ncbi.nlm.nih.gov/29716592/

[4] Ismail IN, Jalaludin J, Bakar SA, Hisamuddin NH, Suhaimi NF 2019 Association of Traffic-Related Air Pollution (TRAP) with DNA Damage and Respiratory Health Symptoms among Primary School Children in Selangor *Asian J. Atmos. Environ* [Internet] [cited 2019 Dec 10];**13**(2):106–16. Available from AJAE: http://asianjae.org/_PR/view/?aidx=20136&bidx=1601#!po=2.08333

[5] Muhammad NS, Jalaludin J, Sundrasegaran S 2014 Exposure to respirable dust (PM10) and

respiratory health among traffic policemen in Selangor. *Adv Environ Biol* [Internet] [cited 2019 Dec 10];**8**(15):199–206. Available from ResearchGate: https://www.researchgate.net/publication/276028026_Exposure_to_respirable_dust_pm10_and_respiratory_health_Among_traffic_policemen_in_selangor

[6] Environment Agency of East Java Province 2018 *Environmental Management Information Report of East Java 2017*. Surabaya

[7] US EPA 2019 *Sulfur Dioxide Basics | Sulfur Dioxide (SO_2) Pollution* [Internet]. United States: [cited 2109 Dec 15]. Available from: https://www.epa.gov/SO_2 -pollution/sulfur-dioxide-basics

[8] US EPA 2009 *Primary National Ambient Air Quality Standard for Nitrogen Dioxide* [Internet]. Federal Register; 2009 Dec 08 [cited 2019 Dec 15]. Available from: https://www.epa.gov/SO_2 -pollution/primary-national-ambient-air-quality-standard-naaqs-sulfur-dioxide

[9] Zakaria N, Azizah R 2013 Analysis of Air Pollution (SO_2), Subjective complaint of throat irritation and eye irritation complaints at food vendors around the Joyoboyo Surabaya terminal. *Indones Occup Saf dan Heal* [Internet]. [cited 2019 Dec 15];**2**(1):75–81. Available from Journal Unair:http://journal.unair.ac.id/K3@analisis-pencemaran-udara-(SO_2),-keluhan-iritasi-tenggorokan-dan-keluhan-kesehatan-iritasi-mata-pada-pedagang-makanan-di-sekitar-terminal-joyoboyo-surabaya-article-6720-media-39-category-16.html

[10] Altuğ H, Gaga EO, Döğeroğlu T, Brunekreef B, Hoek G, Van Doorn W 2014 Effects of ambient air pollution on respiratory tract complaints and airway inflammation in primary school children. *Sci Total Environ* [Internet] [cited 2019 Dec 15];**479–480**(1):201–9 Available from PubMed: https://pubmed.ncbi.nlm.nih.gov/24561926/

[11] Muhamad Daud SA, Jalaludin J, Sopian NA 2018 Air pollutants exposure and frequency of micronuclei (MN) among primary school children nearby industrial area. *Malaysian J Med Heal Sci* [Internet] [cited 2019 Dec 15];**14**(SP2):56-62. Available from Semantic Scholar: https://www.semanticscholar.org/paper/Air-pollutants-exposure-and-frequency-of-(MN)-among-Daud Jalaludin/74ce133736cdf5033e8b457ab935e3cd4227b172

[12] Health Office of East Java Province 2015 *East Java Health Profile 2015*. Surabaya; 2015.

[13] Health Office of East Java Province 2016 *East Java Health Profile 2016* [Internet]. Surabaya. 2016 [cited 28 August 2020] Available from https://www.kemkes.go.id/resources/download/pusdatin/profil-kesehatan-indonesia/indonesia-health-profile-2016.pdf

[14] Health Office of East Java Province 2017 *East Java Health Profile 2017* [Internet]. Surabaya. 2017 [cited 28 August 2020] Available from https://www.kemkes.go.id/resources/download/pusdatin/profil-kesehatan-indonesia/indonesia-health-profile-2017.pdf

[15] Cai Y, Zhang B, Ke W, Feng B, Lin H, Xiao J, et al 2016 Associations of Short-Term and Long-Term Exposure to Ambient Air Pollutants With Hypertension: A Systematic Review and Meta-Analysis. *Hypertension* 2016;**68**(1):62–70 DOI: 10.1161/HYPERTENSIONAHA.116.07218

[16] Choi JH, Xu QS, Park SY, Kim JH, Hwang SS, Lee KH, et al 2007 Seasonal variation of effect of air pollution on blood pressure *J Epidemiol Community Health* **61**(4):314–8. DOI: 10.1136/jech.2006.049205

[17] Environment Agency of East Java Province 2016 *Main Report* [Internet]. Surabaya. 2016 [cited 20 August 2020] Available from https://www.env.go.jp/earth/coop/lowcarbon-asia/english/project/data/02eng_IDN_H27_02.pdf

[18] Health Office of East Java Province 2016 *East Java Health Profile 2016* [Internet]. Surabaya. 2016 [cited 28 August 2020] Available from https://www.kemkes.go.id/resources/download/pusdatin/profil-kesehatan-indonesia/indonesia-health-profile-2016.pdf

[19] Environment Agency of East Java Province 2016 *Environmental Management Information Report of East Java 2016* [Internet]. Surabaya. 2017 [cited 21 August 2020] Available from https://www.adb.org/sites/default/files/publication/527406/governance-brief-037-east-java-

innovation-hub.pdf
[20] Chan C, Yang H, Lin R 2008 A community-based study on the association between hypertension and air pollution *Epidemiology* 2008;**19**(6) : 1–21
[21] Zhang Q, Tian J, Bai Y, Yang Z, Zhang H, Meng Z 2013 Effects of Sulfur Dioxide and Its Derivatives on the Functions of Rat Hearts and their Mechanisms *Procedia Environ Sci* **18**:43–50. doi: 10.1016/j.proenv.2013.04.007
[22] Gunnison AF, Benton AW Sulfur Dioxide: Sulfite 1971 *Arch Environ Heal An Int J* **22**(3):381–8. https://doi.org/10.1080/00039896.1971.10665860
[23] Meng Z, Li Y, Li J 2007 Vasodilatation of sulfur dioxide derivatives and signal transduction *Arch Biochem Biophys* **467**(2):291–6. DOI: 10.1016/j.abb.2007.08.028
[24] Meng Z, Geng H, Bai J, Yan G 2003 Blood pressure of rats lowered by sulfur dioxide and its derivatives *Inhalation Toxicology* **15**(9): 951–9 DOI: 10.1080/08958370390215785
[25] Huang Y, Tang C, Du J, Jin H 2016 Endogenous sulfur dioxide: A new member of gasotransmitter family in the cardiovascular system. *Oxid Med Cell Longev*
[26] Drew RT, Kutzman RS, Costa DL, Iwai J 1983 Effects of sulfur dioxide and ozone on hypertension sensitive and resistant rats Vol. **3** *Toxicological Sciences* p. 298–302.
[27] Lu W, Sun Y, Tang C, Ochs T, Qi J, Du J, et al 2012 Sulfur dioxide derivatives improve the vasorelaxation in the spontaneously hypertensive rat by enhancing the vasorelaxant response to Nitric oxide. Vol. **237**, *Experimental Biology and Medicine* p. 867–72.
[28] Scaife A, Barclay J, Hillis GS, Srinivasan J, Macdonald DW, Ross JAS, et al 2012 Lack of effect of nitrogen dioxide exposure on heart rate variability in patients with stable coronary heart disease and impaired left ventricular systolic function *Occupational and Environmental Medicine* **69**(8):587–91.
[29] Langrish JP, Lundbäck M, Barath S, Söderberg S, Mills NL, Newby DE, et al 2010 Exposure to nitrogen dioxide is not associated with vascular dysfunction in man Vol. **22** *Inhalation toxicology* p. 192–8.
[30] Coogan PF, White LF, Yu J, Brook RD, Burnett RT, Marshall JD, et al 2017 Long-term exposure to NO_2 and ozone and hypertension incidence in the black women's health study *Am J Hypertens* **30**(4):367–72.
[31] Howell NA, Tu J V, Moineddin R, Chen H, Chu A, Hystad P, et al 2019 Interaction between neighborhood walkability and traffic-related air pollution on hypertension and diabetes : The CANHEART cohort. *Environ Int* **132**(January):104799.
[32] Molitor J, Jerrett M, Chang C, Molitor N, Gauderman J, Berhane K, et al 2007 Assessing Uncertainty in Spatial Exposure Models for Air Pollution Health Effects Assessment *Environ Health Perspect* Aug; **115**(8): 1147–1153.
[33] Department of the Environment and Energy Australia 2005 *Sulfur dioxide (SO_2) - Air quality fact sheet.* Australian Government. Canberra: Department of the Environment and Energy
[34] World Health Organization 2018 *Ambient (outdoor) air pollution* Geneva: World Health Organization
[35] Omar D, Nazli SN, Karuppannan SA 2012 Clinical Waste Management in District Hospitals of Tumpat, Batu Pahat and Taiping *Procedia - Soc Behav Sci* **68** :134–45.
[36] Azizah R, Eriyani ID, Chaigarun S 2018 Management of Solid Medical Waste in Hospital a Logic Model Approach : A Literature Review *Malaysian J. Med Health Sci* **15**:2018–20.

Evaluation of In-Vehicle Pollutants Exposure during Movement Control Order (MCO) and Respiratory Symptoms among Bus Drivers in Kota Bahru, Malaysia

Zhafir Adam Ayub[1], Juliana Jalaludin[1]*, Khairul Nizam Mohd Isa[1,2] and Nor Mohd Razif Noraini[3]

[1] Department of Environmental and Occupational Health, Faculty of Medicine and Health Sciences, Universiti Putra Malaysia, UPM, Serdang, Selangor, Malaysia.
[2] Environmental Health Research Cluster (EHRc), Environmental Healthcare Section, Institute of Medical Science Technology, Universiti Kuala Lumpur, Kajang, Selangor, Malaysia.
[3] Consultation, Research and Development Department, National Institute of Occupational Safety and Health (NIOSH), Bangi, Malaysia

* Corresponding author's email: juliana@upm.edu.my

Abstract. Exposure to in-vehicle air pollutants has been recognized as a risk factor for respiratory diseases. The present study therefore investigates the factors, including duration of employment and in-vehicle concentrations of fine particulate matter ($PM_{2.5}$), carbon dioxide (CO_2) and nitrogen dioxide (NO_2) that can influence the prevalence of respiratory symptoms in commercial bus drivers. A total of 34 bus drivers and 51 administrative staff (comparative group) from the bus operator in Kota Bahru, Malaysia were randomly enrolled in this cross-sectional study. The personal characteristics and respiratory symptoms were reported using a validated questionnaire from the American Thoracic Society (ATS-DLD-78-A). In-vehicle concentrations of $PM_{2.5}$, CO_2 and NO_2 were continuously measured over eight hours on each bus driver. The prevalence of self-reported cough, phlegm and chronic phlegm among bus drivers were 55.9%, 73.5% and 35.3%, respectively, significantly higher compared to the comparative group ($p < 0.001$). Moreover, the duration of employment with bus operators for more than four years was significantly related to the manifestation of phlegm (OR = 8.27, 95% CI = 3.12-21.89) and cough (OR = 4.57, 95% CI = 1.66-12.59). Among the pollutants measured, concentrations of $PM_{2.5}$ (92.1 $\mu g/m^3$) and CO_2 (486.8 ppm) inside the buses were significantly higher in the morning compared to afternoon hours ($p < 0.001$). These findings provide important evidence for further surveillance programs that could reduce the occupational exposure for bus drivers.
Keywords: Bus driver, In-vehicle pollutants, $PM_{2.5}$, CO_2, Respiratory symptoms

1. Introduction

A considerable amount of epidemiological studies have linked in-vehicle air pollutants, including particulate matters (PM), nitrogen dioxide (NO_2), ozone (O_3) and sulphur dioxide (SO_2) with respiratory morbidity and mortality, particularly asthma and chronic obstructive pulmonary disease (COPD) [1,2]. Taxi drivers, bus drivers, truck drivers as well as passengers are commonly exposed to the in-vehicle air pollutants in the course of their work or travel times. Additionally, they represent a substantial proportion of workforce in urban areas [3]. Furthermore, the concentration of air pollutants in urban areas increased proportionally to the density of vehicles on the road, which can be infiltrated into the vehicles [4,5].

More recently, two reviews found associations between longer average periods of exposure to in-vehicle pollutants and reduction of respiratory performance, airway inflammation and manifestation of respiratory symptoms among commercial drivers [1,6]. Nevertheless, there is scarcity of evidence-based data on the assessment of the effects of the in-vehicle pollutants on bus drivers. Thus, it becomes necessary to consider this problem in a more thorough way. For these reasons, we conducted the present study to determine level of in-vehicle concentrations of $PM_{2.5}$, CO_2, and NO_2 during the Movement Control Order (MCO) period and to explore association of duration of employment with the prevalence of respiratory symptoms in commercial bus drivers in Kota Bahru, Malaysia.

2. Material and Methods

2.1. Study Location and Population

This cross-sectional study was carried out in Kota Bahru City, Kelantan, Malaysia, which is experiencing rapid urbanization. Based on a 2010 census data, Kota Bahru has a population of 44,757 and a size of 394 km^2 [7]. A total of 85 respondents, including 34 bus drivers (exposed group) and 51 employees from the administrative department (comparative group) were enrolled in this study. They were randomly selected from the public bus operators in Kota Bahru.

The inclusion criteria of the respondents were restricted to male drivers, between 20 and 56 years old who had been working with bus operators for at least one year prior to this study. The employees with past history of severe underlying diseases were excluded. This study was conducted from February 2021 to March 2021.

The study protocol was approved by the Ethic Committee for Research Involving Human Subjects in Universiti Putra Malaysia (JKEUPM-2021-027), and obtained all necessary approvals from bus operators. Written informed consent was obtained from all individual participants prior to the study.

2.2. Health Effect Monitoring and Exposure Assessment

A validated American Thoracic Society (ATS-DLD-78-A) questionnaire was self-administered in the native language by the participants to obtain information on personal characteristics and respiratory symptoms. We focused on four different respiratory symptoms: cough, phlegm, wheeze and breathlessness, which were treated as outcome variables. Then, a quick face-to-face interview was carried out to verify the information in the questionnaire.

The occupational exposure to $PM_{2.5}$, NO_2 and CO_2 were recorded with DustTrak™ II Aerosol Monitor (Model 8532 TSI Incorporated, Minnesota, USA), Aeroqual Series 500 (Aeroqual, Auckland, New Zealand) and TSI Q-TRAK Indoor Air Quality Monitor (Model 7565 TSI Incorporated, Minnesota, USA), respectively. These devices were calibrated and secured near the bus driver's seat. All measurements were continuously recorded for two time slots in a day, comprising of the morning hours (7.00 AM-11.00 AM) and afternoon hours (1.00 PM-5.00 PM), to understand the impact of traffic on the levels of $PM_{2.5}$, NO_2 and CO_2. All the respondents were monitored once during the eight hours of work as previously described [8]. The pollutant monitoring was deployed in 22 different commercial buses, which have a capacity for 40 passengers, powered by a diesel engine and equipped with an air-conditioning system. All buses used different urban routes networks in Kota Bahru, but had similar traffic conditions.

2.3. Statistical Analysis

Statistical analyses were performed using the Statistical Package for Social Science (SPSS) 25.0. The Chi-square test was used to evaluate the association and comparison between two groups. A Mann-Whitney test was carried out to evaluate the differences of pollutant concentrations between morning and afternoon hours. Analyses were two-sided and p less than 0.05 were considered statistically significant.

3. Results

3.1. Characteristics and Prevalence of Respiratory Symptoms of Studied Groups

The personal characteristics obtained from the questionnaire are summarized in Table 1. All of the participants were male. The majority of the respondents for both groups were between 32 and 60 years old ($p < 0.001$). Current smokers represented 73.5% and 62.7% of bus drivers and administrative staff, respectively ($p > 0.005$). Furthermore, more than half of the bus drivers had education beyond high school level ($p < 0.05$).

The prevalence of cough, phlegm and chronic phlegm were significantly higher in bus drivers group (55.9%, 73.5% and 35.3%, respectively) compared to comparative group (13.7%, 35.3% and 3.9%, respectively) ($p < 0.001$) (Table 2).

Table 1. Personal characteristics of bus drivers (n = 34) and administrative staff (n = 51).

Characteristics		Bus drivers n (%)	Administrative staff n (%)	p-value
Age	20 to 31 (year)	6 (17.6)	30 (58.8)	< 0.001**
	32 to 60 (year)	28 (82.4)	21 (41.2)	
Education level	Low	25 (73.5)	21 (41.2)	0.003*
	High	9 (26.5)	30 (58.8)	
Duration of employment	≤ 4 years	1 (2.9)	44 (86.3)	< 0.001**
	> 4 years	33 (97.1)	7 (13.7)	
Smoking status	Yes	25 (73.5)	32 (62.7)	0.300
	No	9 (26.5)	19 (37.3)	

* $p < 0.05$; ** $p < 0.001$

Table 2. Prevalence of respiratory symptoms of the studied groups.

Respiratory symptom		Bus drivers n (%)	Administrative staff n (%)	p-value
Cough	Yes	19 (55.9)	7 (13.7)	< 0.001**
	No	15 (44.1)	44 (86.3)	
Chronic cough	Yes	2 (5.9)	1 (2.0)	0.343
	No	32 (94.1)	50 (98.0)	
Phlegm	Yes	25 (73.5)	18 (35.3)	0.010*
	No	9 (26.5)	33 (64.7)	
Chronic phlegm	Yes	12 (35.3)	2 (3.9)	< 0.001**
	No	22 (64.7)	49 (96.1)	
Wheezing	Yes	2 (5.9)	0 (0)	0.053
	No	32 (94.1)	51 (100)	
Breathlessness	Yes	1 (2.9)	0 (0)	0.174
	No	33 (97.1)	51 (100)	

* $p < 0.05$; ** $p < 0.001$

3.2. Trend of Pollutants Measured inside the Buses

The observed average concentrations of $PM_{2.5}$ and CO_2 inside the bus for five days in the morning were significantly higher compared afternoon hours (p < 0.001) (Figure 1). Similarly, distinct patterns of $PM_{2.5}$ and CO_2 were observed in the day time series (Table 3). Concentration of $PM_{2.5}$ peaked on day two, while CO_2 peaked on day one with the median values of 109 $\mu g/m^3$ (IQR = 46.75) and 554.5 ppm (IQR = 75.0), respectively. However, the concentrations of NO_2 measured inside the bus were found to be very less and not statistically different between morning and afternoon hours (p > 0.05).

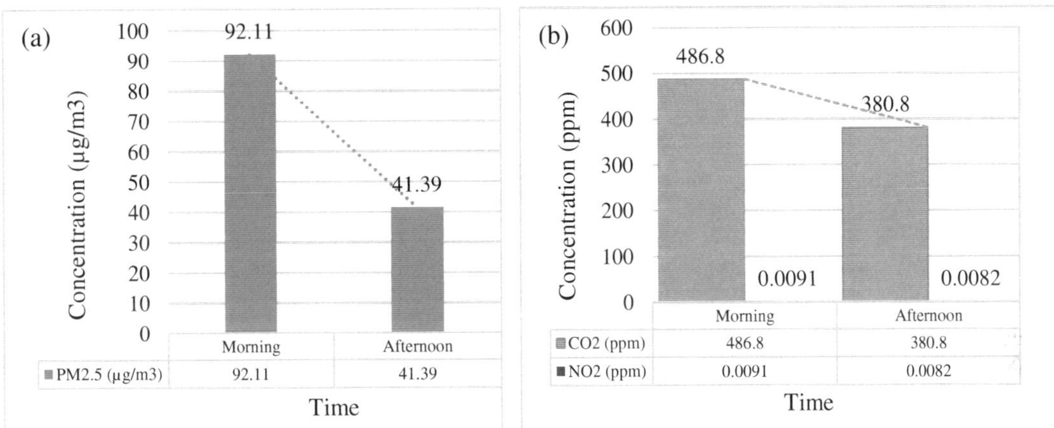

Figure 1. Comparison of average concentration of (a) $PM_{2.5}$ and (b) CO_2 and NO_2 between morning and afternoon hours.

Table 3. Comparison of pollutants in the morning and afternoon hours.

Day	Pollutant	Morning Median (IQR)	Afternoon Median (IQR)	p-value
Day 1	$PM_{2.5}$ ($\mu g/m^3$)	73.0 (36.50)	34.50 (8.75)	< 0.001**
	CO_2 (ppm)	554.5 (75.0)	368.0 (82.0)	< 0.001**
	NO_2 (ppm)	0.001 (0.014)	0.008 (0.01)	0.317
Day 2	$PM_{2.5}$ ($\mu g/m^3$)	109.0 (46.75)	41.0 (9.75)	< 0.001**
	CO_2 (ppm)	522.5 (54.0)	369.5 (38.0)	< 0.001**
	NO_2 (ppm)	0.006 (0.016)	0.01 (0.008)	0.577
Day 3	$PM_{2.5}$ ($\mu g/m^3$)	69.50 (26.25)	38.50 (14.50)	< 0.001**
	CO_2 (ppm)	404.0 (59.0)	382.0 (51.0)	< 0.001**
	NO_2 (ppm)	0.00 (0.008)	0.003 (0.005)	0.218
Day 4	$PM_{2.5}$ ($\mu g/m^3$)	88.50 (30.50)	39.0 (13.25)	< 0.001**
	CO_2 (ppm)	481.0 (83.0)	390.0 (52.0)	< 0.001**
	NO_2 (ppm)	0.004 (0.015)	0.009 (0.008)	0.353
Day 5	$PM_{2.5}$ ($\mu g/m^3$)	88.50 (55.0)	40.50 (17.25)	< 0.001**
	CO_2 (ppm)	472.0 (38.0)	394.50 (26.0)	< 0.001**
	NO_2 (ppm)	0.007 (0.015)	0.005 (0.005)	0.786

** $p < 0.001$
IQR = Interquartile range

3.3. Association between Duration of Employment and Respiratory Symptoms

Table 4 lists the prevalence and odd ratio for respiratory symptoms in relation to the duration of employment. Overall, respiratory symptoms were more common among respondents employed for more than four years than for those with a shorter duration of employment with bus operators. In particular, the risk of phlegm occurrence was the highest among the four respiratory symptoms (OR = 8.27, 95% CI = 3.12-21.89), followed by manifestation of cough with an odd of 4.57 (95% CI= 1.66-12.59).

Table 4. Prevalence of respiratory symptoms by duration of employment (N= 85).

Respiratory symptom		Duration of employment ≤ 4 years n (%)	Duration of employment > 4 years n (%)	p-value	OR	95% CI
Cough	Yes	7 (15.9)	19 (46.3)	0.002*	4.57	1.66-12.59
	No	37 (84.1)	22 (53.7)			
Chronic cough	Yes	1 (2.3)	2 (4.9)	0.512	2.21	0.19-25.28
	No	43 (97.7)	39 (95.1)			
Phlegm	Yes	12 (27.3)	31 (75.6)	< 0.001**	8.27	3.12-21.89
	No	32 (72.7)	10 (24.4)			
Chronic phlegm	Yes	0 (0)	14 (34.1)	< 0.001**	-	-
	No	44 (100)	27 (65.9)			
Wheezing	Yes	0 (0)	2 (4.9)	0.085	-	-
	No	44 (100)	39 (95.1)			
Breathlessness	Yes	0 (0)	1 (2.4)	0.225	-	-
	No	44 (100)	40 (97.6)			

* $p < 0.05$; ** $p < 0.001$
OR = Odd ratio ; CI = Confidence interval

4. Discussion

This study demonstrated that manifestation of respiratory symptoms of cough, phlegm and chronic phlegm were more common among bus drivers than administrative employees. This similar pattern of prevalence was observed in previous studies conducted in Kuala Lumpur, Malaysia [5], City of Dakar, Senegal [9] and Bursa City, Turkey [10]. Review reported that occupational exposure to air pollution among commercial drivers is linked to adverse health outcomes [3]. Moreover, a high proportion of active smokers (73.5%) among bus drivers was observed in this current study and those reported by Mbelambela et al. [11], where they found that 71.0% of bus driver in Kinshasa, Republic of Congo were active smokers.

Furthermore, this study provides evidence in line with previous epidemiology studies that bus drivers employed for more than four years demonstrated a higher prevalence of respiratory symptoms than those with shorter employment history [12]. Additionally, there are two previous studies that discovered both smokers and non-smokers bus drivers with longer employment history had a higher risk of COPD. They found that the duration of employment of more than 10 years and 20 years exhibited higher prevalence of respiratory symptoms among bus drivers in Hyderabad City, India and urban areas of North Macedonia, respectively [13,14]. Taken together, these findings suggest that bus drivers are experiencing a cumulative effect of occupational exposure.

This current study confirms that bus drivers are exposed to very high levels of $PM_{2.5}$ while at work, especially in the morning. In five days of monitoring, the concentrations of $PM_{2.5}$ were exceeding the 24 hours mean of World Health Organization (WHO) guideline (25 $\mu g/m^3$), the National Ambient Air Quality Standard by U.S. EPA (35 $\mu g/m^3$) and the new Malaysian Ambient Air Quality Standard 2018 Interim Target-2 (50 $\mu g/m^3$). Up to the present time, there is no particular standard level for $PM_{2.5}$ set by the Occupational Safety and Health Act (OSHA, Act 514 1994) or other work exposure standards for Time Weighted Average (TWA) [15]. Therefore, $PM_{2.5}$ exposure level cannot be compared directly with OSHA standard. These findings further support the results from previous studies that in-vehicle concentrations of particulate matter were often higher compared with outdoor micro-environment, which originate from external sources including exhaust emissions [16,17]. One study from Istanbul city, Turkey reported that the concentration of fine and course particles inside buses and cars were greater during rush hours than those during off-peak hours, mostly caused by exhaust emission and resuspension of particles [18].

Our results demonstrated that the NO_2 concentrations inside the buses measured for five days were considerably lower than the standards set by the WHO guidelines (200 $\mu g/m^3$) [19] and the National Ambient Air Quality Standard by U.S. EPA (100 ppb) [20]. Moreover, NO_2 concentrations inside the buses reported from various countries were in the range between 15 ppb and 117 ppb [21,22], which is significantly higher compared to our own result. Several factors are known to be partially responsible for this variation. Many studies have demonstrated that the variation of NO_2 concentrations in-vehicle were significantly influenced by idle/close ventilation conditions, ambient temperature, time of the day, seasons, road traffic patterns, filters, fuel type, vehicle age, level of exhaust emissions and outdoor NO_2 levels [23–25]. Nevertheless, our study has been unable to demonstrate the differences of NO_2 concentration between morning and afternoon hours. This result may be explained by the fact that this study was performed during the Movement Control Order (MCO) to reduce the number of coronavirus 2 (SARS-CoV-2) infections, which caused low traffic intensity and eventually reduced the concentrations of air pollutants, including NO_2, CO, CO_2, PM_{10} and $PM_{2.5}$. This is supported by a recent study conducted in Klang Valley, Malaysia during the MCO by Latif et al. [26], who identified that the air concentrations of NO_2 were reduced between -55% and -72% compared with the concentrations from the same periods in 2019 and 2018.

In this current study, the average concentrations of CO_2 ranged from 368.0 to 554.5 ppm, therefore they were much lower than the recommended limit of 1,000 ppm set by the WHO guidelines and the Industrial Code of Practice on Indoor Air Quality (ICOP-IAQ) 2010. This could be explained by the fact that mechanical air circulation inside the vehicle is able to reduce cabin air pollutants [6]. Furthermore, the concentration of CO_2 in-vehicle is also proportional to the number of passengers as demonstrated in a study conducted by Li et al. [27] using CO_2 sensor MH-Z16 and machine learning analysis. More recently, one study conducted using the public bus system in Barcelona, NE Spain reported that in-vehicle CO_2 concentration not only depends on occupancy ratios, but also travelling time, speed of vehicle, ventilation system and opening of the doors and trapdoors on the roof [28]. Another factor that has a significant impact on the concentration of CO_2 is linked with MCO implementation. A recent review reported that the CO_2 emissions in Malaysia were decreased significantly by 9.7% from 250.09 $MtCO_2$ in 2019 to 225.97 $MtCO_2$ in 2020 due to stringent MCO implementation [29].

There are several limitations in this study. First, the nature of cross-sectional study design did not allow the determination of causal relationship. Second, the information on the respiratory symptoms was obtained from a self-administrated questionnaire, which was subject to recall bias. Nevertheless, the impact of this bias was controlled with verification procedures through face-to-face interview after completing the questionnaire. Third, the bus occupancy and traffic density were lower than usual, due to implementation of the MCO and the fear of SARS-CoV-2 infection. Finally, other factors such as different bus travelling routes, ambient air pollution and meteorological variables could have biased our results.

5. Conclusion

In conclusion, our findings showed that the manifestation of cough and phlegm were more common among respondents employed for more than four years than for those with a shorter history of employment with bus operators. Furthermore, the average concentration of CO_2 and NO_2 were below the recommended limits set by the WHO, and the National Ambient Air Quality Standard by U.S. EPA, except for the concentration of $PM_{2.5}$. The diurnal patterns demonstrated an increase in all measured pollutants except NO_2 in the morning hours. This study suggests further longitudinal study design to examine the long-term effects of exposure to in-vehicle pollutants and surveillance programs to reduce the effects of occupational exposure for bus drivers.

Acknowledgement
We are grateful to all participants that took part in this work.

References

[1] Crizzle AM, Bigelow P, Adams D, Gooderham S, Myers AM, Thiffaul P 2017 Health and wellness of long-haul truck and bus drivers: A systematic literature review and directions for future research *J Transp Heal J* **7A** 90–109
[2] Brucker N, Moro AM, Charão MF, Durgante J, Freitas F, Baierle M, Nascimento S, Gauer B, Bulcão RP, Bubols GB, et al. 2013 Environment Biomarkers of occupational exposure to air pollution, inflammation and oxidative damage in taxi drivers *Sci Total Environ* **463–464** 884–893
[3] Lawin H, Fanou LA, Hinson AV, Stolbrink M, Houngbegnon P, Kedote NM, Fayomi B, Kagima C, Katoto P, et al. 2018 Health Risks Associated with Occupational Exposure to Ambient Air Pollution in Commercial Drivers: A Systematic Review *Int J Environ Res Public Health* **15(2039)** 1–13
[4] Azhari A, Latif MT, Mohamed AF 2018 Road traffic as an air pollutant contributor within an industrial park environment *Atmos Pollut Res* **9(4)** 680–687
[5] Mohd Firdaus O, Jalaludin J 2014 Exposure to Indoor Air Pollutants (PM10, CO2 And CO) and Respiratory Health Effects among Long Distance Express Bus Drivers *Heal Environ J* **5(3)** 66–85
[6] Brucker N, Nascimento SN do, Bernardini L, Charao MF, Garcia SC 2020 Biomarkers of exposure, effect, and susceptibility in occupational exposure to traffic-related air pollution: A review *J Appl Toxicol* **40(6)** 722–736
[7] Department of Statistics Malaysia (DOSM). Current Population Estimates, Malaysia, 2020 https://www.dosm.gov.my
[8] Kavitha M, Jalaludin J, Addah MA 2011 Relationship Between Exposure To Particulate Matter And Biomarkers Among Bus Drivers In Klang Valley, Malaysia *Heal Environ J* **2(2)** 1–7
[9] Sylla FK, Faye A, Diaw M, Fall M, Tal-Dia A 2018 Traffic Air Pollution and Respiratory Health: A Cross-Sectional Study among Bus Drivers in Dakar (Senegal) *Open J Epidemiol* **8** 1–13
[10] Dikiş ÖŞ, Yildiz T, Dulger SU, Kacan CY, Haberal MA, Cetin T 2020 Does occupational air pollution threaten the lung health of indoor workers more than those of bus drivers?: a cross-sectional study *Aging Male* **23(5)** 934–40
[11] Mbelambela EP, Hirota R, Eitoku M, Muchanga SMJ, Kiyosawa H, Yasumitsu-Lovell K, Suganuma S 2017 Occupation exposed to road-traffic emissions and respiratory health among Congolese transit workers, particularly bus conductors, in Kinshasa: a cross-sectional study *Environ Health Prev Med* **22(11)** 1–9
[12] Zuskin E, Mustajbegovic J, Schachter N 1994 Respiratory Symptoms and Lung Function in Bus Drivers and Mechanics *Am J Ind Med* **26(6)** 771–83
[13] Minov J, Stoleski S, Vasilevska K, Mijakoski D, Atanasovska A, Bislimovska D 2019 Chronic obstructive pulmonary disease in urban bus drivers *Eur Respir J* **54(Suppl.63)** PA2837
[14] Faisal M, Chelluri PE, Singaraju S, Gali JH, Ahmed S, Srinivas P 2013 Environmental and occupational respiratory diseases – 1063. Spirometric abnormalities in non smoking bus drivers of hyderabad *World Allergy Organ J* **6(Suppl 1)** 61
[15] Jamil PASM, Karuppiah K, Rasdi I, How V, Tamrin SBM, Mani KKC, Sambasivam S, Yusof NADM, Hashim NM 2020 Associations of Occupational, Socio-Demographic and Lifestyle Factors with Lung Functions in Malaysian Traffic Policemen *Ann Glob Heal* **58(1)** 84
[16] Gany F, Bari S, Prasad L, Leng J, Lee T, Thurston GD, Gordon T, Acharya S, Zelikoff JT 2017 Perception and reality of particulate matter exposure in New York City taxi drivers *J Expo Sci Environ Epidemiol* **27** 221–226

[17] Geiss O, Barrero-moreno J, Tirendi S, Kotzias D 2010 Exposure to Particulate Matter in Vehicle Cabins of Private Cars *Aerosol Air Qual Res* **10** 581–588

[18] Onat B, Stakeeva B 2013 Personal exposure of commuters in public transport to PM2.5 and fine particle counts *Atmos Pollut Res* **4(3)** 329–335

[19] World Health Organization 2018 Ambient (outdoor) air pollution https://www.who.int/news-room/fact-sheets/detail/ambient-(outdoor)-air-quality-and-health

[20] Environmental Protection Agency 2021 NAAQS https://www.epa.gov/criteria-air-pollutants/naaqs-table

[21] Zak M, Melaniuk-Wolny E, Widziewicz K 2017 The exposure of pedestrians, drivers and road transport passengers to nitrogen dioxide *Atmos Pollut Res* **8** 781–790

[22] Heberle SM, Lorini C, Rosa MSG, Barros N 2019 Evaluation of bus driver exposure to nitrogen dioxide levels during working hours *Atmos Environ* **216** 1–9

[23] Kadiyala A, Kumar A 2012 An Examination of the Sensitivity of Sulfur Dioxide, Nitric Oxide, and Nitrogen Dioxide Concentrations to the Important Factors Affecting Air Quality Inside a Public Transportation Bus *Atmosphere* **3** 266–287

[24] Mohtar AAA, Latif MT, Baharudin NH, Ahamad F, Chung JX, Othman M, Juneng L 2018 Variation of major air pollutants in different seasonal conditions in an urban environment in Malaysia *Geosci Lett* **5(21)** 1–13

[25] Lim S, Holliday L, Barratt B, Griffiths CJ, Mudway IS 2021 Assessing the exposure and hazard of diesel exhaust in professional drivers: a review of the current state of knowledge *Air Qual Atmos Heal* **14** 1681–1695

[26] Latif MT, Dominick D, Hawari NSSL, Mohtar AAA, Othman M 2020 The Concentration of Major Air Pollutants during the Movement Control Order due to the COVID-19 Pandemic in the Klang Valley Malaysia *Sustain Cities Soc* **66** 102660

[27] Li T, Fong S, Yang L 2018 Counting Passengers in Public Buses by Sensing Carbon Dioxide Concentration: Data Collection and Machine Learning *Proceedings of the 2018 2nd International Conference on Big Data and Internet of Things* (Beijing, China/Association for Computing Machinery) p 43–48

[28] Querol X, Alastuey A, Moreno N, Minguillon MC, Moreno T, Karanasiou A, Jimenez JL, Li Y, Morguí JA, Felisi JM 2022 How can ventilation be improved on public transportation buses? Insights from CO2 measurements *Environ Res* **205** 1–13

[29] Tan CH, Ong MY, Nomanbhay SM, Shamsuddin AH, Show PL 2021 The influence of COVID-19 on glocal CO2 emissions and climate change: A perspective from Malaysia *Sustainability* **13(15)** 1–9

The Analysis of Pollutant Concentrations Before and During Movement Control at Quarry Areas Using an Artificial Intelligent Technique

N Z Yahaya[1*], J Jalaludin[2], I A I Rais [3], U S Mohamad Hussin[4], J Yahaya[5], E S A Rahman[6]

[1]Faculty of Ocean Engineering Technology and Informatics, University Malaysia Terengganu, Kuala Terengganu, 21030, Malaysia
[2]Faculty of Medicine and Health Sciences, Universiti Putra Malaysia, Serdang, Selangor, Malaysia
[3]Pusat Penyelidikan Mineral Malaysia, Jabatan Mineral dan Geosains Malaysia, Ipoh, Perak, Malaysia
[4]Fakulti Pengurusan dan Teknologi Maklumat, Universiti Sultan Azlan Shah, Kuala Kangsar, Perak, Malaysia
[5]Faculty of Information Science and Technology, Universiti Kebangsaan Malaysia, Bangi, Selangor, Malaysia
[6]Department of Environment (Air Division), Ministry of Water and Environment, Putrajaya, Malaysia
*Corresponding author: nzaitun@umt.edu.my

Abstract. This paper investigated the influence and interactions of air pollution concentrations by using the stochastic boosted regression trees between variables for each station and the impact of the COVID-19 Movement Control Order at Ipoh City air quality station. The one-hour data were gathered from the Department of Environment from January until June 2019 and 2020. Two thousand two hundred thirty-one data of particles, gases (Nitrogen oxides, Sulphur Dioxide, Ozone, Carbon Monoxide) concentrations and meteorological data (wind speed, wind directions, temperature, and relative humidity) were captured. The BRT model development process with an algorithm using a comprehensive package, R Software and its packages to understand the variability and trends. It was found that the relationship between the number of samples and number of trees (*nt)* of 4372 for *oob* were found the best iterations obtained. The performance of the boosting model was assessed and found that the FAC2 was 0.91, the R^2 values were above 0.56 (R = 0.74), and the Index of Agreements (IOA) was 0.67, which fall ranges are within an acceptable for model performance. The Relative Variable Importance (RVI) that influenced $PM_{2.5}$ for non-MCO data was CO (18.9%), SO_2 (14.6 %), O_3 (12.9 %), and wd (10.66 %) while CO (22.6%), RH (13.4%), 14.7% and O_3 (12.1%) were RVI factors influenced to PM2.5 concentrations during MCO periods. Estimating the strength of interaction effects (SIE) between variables was 0.24 for CO-wind directions, followed by 0.19 for ozone-wind speeds and 0.15 for NO_2-CO. Results showed that the model developed was within the acceptable range and could be used to understand particles and identify important parameters that influence particle concentrations.

Keywords: Pollutants; Boosted Regression Trees; R-Packages; Relative Variable Importance, Strength of Interactions

IOP Conf. Series: Earth and Environmental Science 1013 (2022) 012011 doi:10.1088/1755-1315/1013/1/012011

1. Introduction

It was reported that a continuous occurrence of an unknown acute respiratory tract infection was first reported in Wuhan City, Hubei Province, China, on 12 December 2019, originating from the Hunan South China Seafood Market [1]. This coronavirus was initially named the 2019-novel coronavirus (2019-nCoV) by the World Health Organization (WHO). But then the WHO renamed this disease the coronavirus disease 2019 (COVID-19). Later after several efforts and studies taken by the WHO and Chinese scientists, this disease spread worldwide. On 11 March 2020, WHO announced the new coronavirus or COVID-19 illness as a "global pandemic" [2]. Concerning that, governments worldwide have taken several steps and actions to control the spreading of COVID-19. The measures include closing schools, industrial activities, restrictions to public transport, houses of worship closure and many other actions.

The COVID-19 pandemic is thought to originate from China by human contact with non-domestic animals. The origin of the COVID-19 virus is believed to have spread from the wild animal market in Wuhan, China [3, 4]. Infectious rates of diseases occur faster than in the past, and the rate of spread increases with increasing human contact. Between January 2020 and the middle of July 2020, COVID-19 infections were detected in more than 13 million worldwide, and 600 thousand people died. According to the World Health Organization [3], there have been 5,313,098 confirmed cases of COVID-19. In Malaysia, COVID-19 was first reported and imported from Wuhan, China, on 25 January 2020 and spread to eight positive patients within six days [5] since the first case. The situation became unfavourable on 11 March 2020 as the spreading of this virus had worsened across the country until then to control the spreading and mortality due to the virus, the Ministry of Health, Malaysia issued a Movement Control Order (MCO), which was then implemented across the country on 18th March 2020. MCO enforcement came under the Prevention and Control of Infectious Diseases Act 1988 and the Police Act 1967 and would help control the spread of the virus. Table 1 shows pollutant standards for the European Union and World Health Organization and Department of Environment (Air Division), Malaysia.

Table 1. Air Quality Standards for the European Union, World Health Organisation Standards and Department of Environment (Air Division), Malaysia

Pollutants/mean	*European Union Air Quality Standards Limit Value ($\mu g/m^3$)	**WHO Guidelines ($\mu g/m^3$)	***MAAQS (2018) -IT2 ($\mu g/m^3$)	US EPA 2003 ($\mu g/m^3$)
$PM_{2.5}$ (24-hr mean)	NA	15	50 $\mu g/m^3$	35 $\mu g/m^3$
$PM_{2.5}$(Annual)	25 $\mu g/m^3$	5 $\mu g/m^3$	25 $\mu g/m^3$	15.0 $\mu g/m^3$
PM_{10} (24-hr)	50 $\mu g/m^3$	45 $\mu g/m^3$	120 $\mu g/m^3$	150 $\mu g/m^3$
PM_{10}(Annual)	40 $\mu g/m^3$	15 $\mu g/m^3$	45 $\mu g/m^3$	NA
O_3 8-hr mean	120 $\mu g/m^3$	[a]100 $\mu g/m^3$	120 $\mu g/m^3$	0.070 ppm
O_3 1-hr mean	NA	[b]60 $\mu g/m^3$	200 $\mu g/m^3$	0.12 ppm
NO_2 (Hourly)	200 $\mu g/m^3$	25 $\mu g/m^3$	300 $\mu g/m^3$	100 ppb
NO_2 (Annual)	40 $\mu g/m^3$	10 $\mu g/m^3$	75 $\mu g/m^3$	53 ppb
SO_2 (Hourly)	350 $\mu g/m^3$	NA	300 $\mu g/m^3$	75 ppb
SO_2 (24-hour)	125 $\mu g/m^3$	40$\mu g/m^3$	90 $\mu g/m^3$	0.14 ppm
CO (1-hr mean)	NA	26 ppm	35 mg/m^3	35 ppm (40 mg/m^3)
CO (8-hr mean)	10 mg/m^3	9 ppm	10 mg/m^3	9 ppm (10 mg/m^3)

*Sources: EU Air Quality Directive (2008/50/EC) https://ec.europa.eu/environment/air/quality/standards.htm),
**WHO,2021, Air Quality Guidelines: Global update
[a] 100 μg/m3, 8-hour daily maximum - 99th percentile, (i.e. 3-4 exceedance days per year)
[b] 60 μg/m3 8-hour mean, peak season - Peak season is defined as an average, so daily maximum 8-hour mean O_3 concentration in the six consecutive months with the highest six-month running average O_3 concentration
*** Malaysia Air Quality Standards [6]

The Department of Environment (DOE) Malaysia reported that air quality levels were low and worse than before the MCO. According to the DOE, the primary pollutants such as Nitrogen dioxide (NO_2), Sulphur dioxide (SO_2), Carbon monoxide (CO), and particulate matter with 2.5-micrometre size at several locations in an urban environment such as Batu Muda, Cheras and Putrajaya were decreased dramatically. It is significantly shown that the reduction of NO_2 is the highest down to 62% lower than before MCO and followed by CO gases. Reducing NO_2 is the highest down to 62% lower than before MCO and followed by CO gases with a maximum of 22% reduction. The main cities in Malaysia also show the same trends whereby NO_2 reduction was highest recorded, followed by SO_2, CO, and particulate matter. It is good to see that air pollution has also decreased. The percentage of stations that recorded "good" air quality readings increased two-fold from 28% to 57% after MCO.

Abdullah *et al.* studied the air quality status during the 2020 Malaysia MCO due to the 2019 novel coronavirus (2019-nCoV) pandemic by comparing $PM_{2.5}$ levels [5]. The study revealed that the changes of fine particulate matter ($PM_{2.5}$) at 68 air quality monitoring stations were found that the $PM_{2.5}$ concentrations showed a high reduction of up to 58.4% during the MCO. Several red zone areas (N41 confirmed COVID-19 cases) had also reduced up to 28.3% in the $PM_{2.5}$ concentrations variation. A study by Latif *et al.* in Klang Valley air quality data in Malaysia indicated that there were significant reductions ($p<0.05$) of PM_{10}, $PM_{2.5}$, NO_2 and CO during the MCO compared with the same periods in 2019 and 2018 [7]. NO_2 recorded the highest percentage of reduction during the MCO with a percentage reduction between the drops of its precursors.

Various statistical tools have been used to analyse $PM_{2.5}$ data used to predict and forecast Malaysia and elsewhere. A study by Marsha and Larkin used a multiple linear regression scheme that relies on the previous day's $PM_{2.5}$ value and fire and smoke-related variables from satellite observations [8]. It was found that the statistical model obtained was able to explain 78% of the variance in daily ground-level $PM_{2.5}$ data during a smoke event in Western United States monitoring stations. A study by Ul-Saufie *et al.* (2013) reported that the regression method is one of the acceptable modelling techniques that has been used to predict PM_{10} concentration [9]. Compared to the traditional least-squares way, using this method, the influence of outliers can be minimised. The most recent studies have attempted to develop powerful computational intelligence models using machine learning algorithms (neural network) conducted by Abdullah *et al.* (2019) to use PM_{10} concentrations in Malaysia air quality data to predict the desired value [5].

Recent studies have demonstrated that a Boosted Regression Tree (BRT) model with stochasticity (SGB) can be used to obtain the best model that can deal with a high-level complexity and large datasets and yield substantial outcomes [10–13]. The boosting technique is applied to 'boost' the model accuracy of any given learning algorithm. Boosting is a general method that can be used, and it was first developed by Friedman in 2001 [14] and then added the stochastic element in the algorithm by Friedman in 2002 [15]. The difference in the stochastic BRT algorithm is that at each iteration, a subsample of the training data is randomly extracted from the complete training data set without being replaced. Furthermore, extensive work was performed by Carslaw and Taylor [10] and Yahaya [11] in the City of Leeds, UK. Their studies were among the first to have examined and used BRTs to analyse air pollution, traffic, and meteorological data intensively. Munir et al. (2021) compared different approaches for assessing the impact of COVID-19 lockdown on urban air quality in the City of Reading, United Kingdom [16]. In their study, they compared different approaches, namely: (1) Sequential approach – comparing pre-lockdown and lockdown periods 2020; (2) Parallel approach – comparing 2019 and 2020 for the equivalent time of the lockdown period; and (3) Machine learning modelling approach as a method to analyse COVID-19 impact to the air quality.

The BRT stochastic procedure [15] as applied to the gradient boosting algorithm is described as follows: Let $\{y_i, x_i\}_i^N$ of known (y,x) values be the entire training data sample and $\{\pi(i)\}_{,i}^{N}$ be a random permutation of the integers {1,.......N}. Then a random subsample of size $\widetilde{N}<N$ is given by $\{y_{\pi(i)}, x_{\pi(i)}\}_i^{\widetilde{N}}$. The SGB algorithm is then:

1. $\hat{f}_0(x) = arg\, arg\, min_{\gamma} \sum_{1=1}^{N} \quad \psi(y_i, \gamma)$ (1)
2. $Form = 1\, toMdo$:
3. $\{\pi(i)\}_1^N$ = rand_perm $\{i\}_1^N$ (2)
4. $\tilde{y}_{im}$= - $[\frac{\partial\varphi(y_{i,}F(x_i))}{\partial F(x_i)}]_{F(x)=F_{m-1}(x)}$, I = 1, $\tilde{N}$ (3)
5. $\{R_{lm}\}_1^L$) = L – terminal node tree ($\{\tilde{y}_{\pi(i)m}, x_{\pi(i)m}\}_1^{\tilde{N}}$ (4)
6. $\gamma_{lm} = arg\, min_{\gamma} \sum_{x_{\pi(i)\in R_{lm}}} \quad \psi(y_{\pi(i)}, F_{m-1}(x_{\pi(i)}) + \gamma)$ (5)
7. F_m (x) = F_{m-1}(x) + $\upsilon . \gamma_{lm}$ 1(x∈ R_{lm}) (6)
8. $endFor$
9. End Algorithm

According to Friedman (2002), the smaller the fraction $f = \tilde{N}/N$, the more the random samples used in successive iterations will differ [15]. Making the value of f smaller reduces the amount of data available to train the base learner at each iteration. This condition causes the variance associated with the individual base learner estimates to increase. Among the outputs of the BRT is the ability to examine RVI and estimate the SIE obtainable between variables. It is expected that the outcomes from this analysis are; First, the RVI in percentage (%) that will rank which variables most influence the concentrations of pollutants at different categories. This information is vital for the manager and planner to understand which pollutants need attention, especially during strategy planning, in mitigating the pollutant levels taking into place; Second, to estimate the SIE on a scale of 0 – 1, which will explore the interaction of variables.

2. Methods

Pakar Scieno TW Sdn Bhd recorded the data of the major air pollutants such as PM_{10}, $PM_{2.5}$, SO_2, NO_2, CO, O_3 on an hourly basis at the Continuous Air Quality Monitoring Station (CAQMS) by the orders of the Malaysian DOE. The hourly dataset of air pollution from the 1 March 2019 (before MCO) to – 1 July 2020 (during MCO) was obtained from DOE for this study. The data was gathered from the DOE from March until June in 2019 and the same period in 2020 to determine the variability of particulate matter ($PM_{2.5}$ and PM_{10}) and gases such as Sulphur dioxide (SO_2), Nitrogen dioxide (NO_2) and other gases. The data were compared between the selected areas based on COVID-19 most affected areas along with meteorological variables analysis. From the zoning map, the sampling point falls into Institution and Community Facilities. Two DOE monitoring stations were selected which are SMK Jalan Tasek (4° 37' 50.0448", 101° 7' 4.224") and SK Jalan Pegoh (4° 33' 16.1058", 101° 4' 54.192"). The stations are marked as blue in Figure 1.

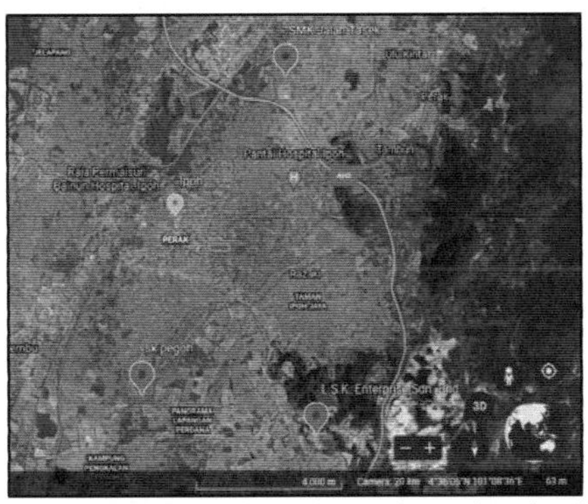

Figure 1. Location of CAQM station at Pegoh and Tasek station in City of Ipoh, Perak, Malaysia (www.google.com/maps/)

3. Results and Discussion

The current famous software used for air quality data analysis is open-source R-package. It was developed by the Development Core Group R (2008). It was used to run and analyse the data obtained from DOE using an extended R Software version 4.1.0 [17] and Openair packages [18, 19]. The list of variables involved in the algorithm setting is shown in Table 2. R is free software, robust and interpreted language with powerful interactive analytical features, making it ideal for quickly developing statistical and data analysis applications [19]. The generalised boosting machine (gbm) package in R was utilised to study the data and produce the BRT algorithms. It was found that the relationship between the number of samples and number of trees (*nt)* of 4372 for *oob* were found the best iterations obtained. The performance of the boosting model was assessed and found that the FAC2 was 0.91, the R^2 values were above 0.56 (R = 0.74), and the Index of Agreements (IOA) was 0.67, which fall ranges are within an acceptable for model performance.

Table 2. Variables names and descriptions used to model boosting algorithm setting

Response Name		Variable names and descriptions	Units
1.	$PM_{2.5}$	Particulate Matter	($\mu g/m^3$)
Explanatory Variables			
1.	O_3	Ozone	(ppm)
2.	NO	Nitric Oxide	(ppm)
3.	NO_2	Nitrogen Dioxide	(ppm)
4.	NO_x	Nitrogen Oxides	(ppm)
5.	SO_2	Sulphur Dioxide	(ppm)
6.	CO	Carbon Monoxide	(ppm)
7.	temp	Ambient Temperature	(^oC)
8.	humid	Relative Humidity	Percentage (%)
9.	ws	Wind Direction and Wind Speed at 10m height	(km/hr)
10.	wd	Wind Direction	(^{o})

The objective here was to find the best combination of parameters (*lr, tc*, and *nt*) that minimised the predictive error (minimal error for predictions to independent samples). The stochastic gradient machine technique was also applied in the BRT model; this technique introduces randomness into the boosted model, typically increases the accuracy and speed, and reduces overfitting [15]. At this stage, the sequential model-fitting process builds on fitted trees. A series of BRT analyses using the gradient boosted model (*GBM* in different settings was performed to examine the *GBM* before determining *tc*, *nt*, and *lr*. The GBM uses three methods to estimate the optimal number of iterations after the GBM model was fitted to identify the optimal number of trees. An independent test set (*test*) method, out-of-bag estimation (*OOB*), and v-fold cross-validation optimisation (CV). Similar to Friedman's MART software [20, 21], it determines the optimal number of iterations by using the independent test set method involving the use of a single holdout base dataset. BRT models with different *numbers of trees* (from 1000 to 10000) were simulated for datasets with *nt* = 10000, *lr* = 0.001, independent depth = 5, and CV fold = 10. A similar shrinkage value (*lr*) of 0.001 was suggested by Ridgeway [20, 21] and used by Carslaw and Taylor [10] and Yahaya [11]for analysing an air pollution dataset.

3.1. BRT Outputs

3.1.1. Relative Variable Importance (RVI)

The Relative Variable Importance (RVI) of the predictor variables was determined using the formulae developed by [14] and implemented in the *gbm* library to estimate the relative importance of predictor variables. At each split in each tree, *gbm* computes the improvement in the split-criterion (MSE for regression). In these methods, the measure is calculated based on the number of times a variable is selected. The effort is used for splitting and is weighted by the improvement in the overall model. The variables with the most significant average decrease in Mean Square Error (MSE) are considered most important. The relative importance analysis focuses on the essential variables obtained from the *gbm* iteration, particles, gases and meteorological variables to the formation of $PM_{2.5}$ in this study. Table 3 shows a summary of relative variable importance (RVI) comparing non-MCO (2019) and MCO situations (2020) data. It indicates that CO (18.9 %) gases are the most dominant gases followed by SO_2 (14.6 %) of RVI values before during non-MCO period influenced $PM_{2.5}$ concentrations followed by meteorological parameters (relative humidity and wind directions). Similar during MCO, CO remained the highest variables to affect $PM_{2.5}$ with 22.6 % RVI, followed by O_3 with 12.1 % and meteorological parameters (temperature and relative humidity). These variables have also been commonly used by other air pollution studies and are known to influence air pollution data, especially during MCO and Non-MCO conditions. The highest differences of RVI are shown by SO_2 levels, which reduced from 14.6% during non-MCO to only 5.7 % during MCO periods. The RVI for O_3 and NO are not so different to give an impact during the pandemic.

Table 3. Relative Variable Importance (RVI) in percentage (%) comparison between 2019 and 2020 (MCO) data for Pegoh station in Ipoh, Perak City

Pollutant	CO	SO_2	NO_x	NO	NO_2	O_3	Temp	RH	WD	WS
2019	18.9	14.6	6.61	6.00	5.11	12.9	9.75	10.66	10.54	4.82
2020	22.6	5.7	3.4	6.6	8.2	12.1	14.7	13.4	7.74	5.6
Differ	+3.7	-8.9	-3.21	+0.6	-3.09	-0.8	+4.95	+2.74	-2.8	+0.78

3.1.2. Strength of Interaction Effects (SIE) and Interaction Indexes

The Strength of Interactions Effects (SIE) between parameters is another vital output in BRT that can explain the level of interactions between variables. De'ath (2007) described how two-way interactions could be used to determine the relative importance of considering the interaction between two variables compared with the purely additive combination [22]. For instance, the two-way interaction of two variables, x_1 and x_2, made predictions across regular interval combinations while keeping all other variables at their mean level. The variable interactions can be determined in BRT by varying the number of splits (size) of the individual trees, determining the degree to which predictors interact in determining the response [11]. The variable interactions can be determined in BRT by varying the number of splits (size) of the individual trees, determining the degree to which predictors interact in determining the response. These functions implement bivariate interpolation onto a grid for irregularly spaced input data. Bilinear or bicubic spline interpolation is applied using different versions of algorithms from *Akima* R-package [23, 24].

Table 4 shows the strength of the interaction index between variables to the level of $PM_{2.5}$ at the station. It can be seen that carbon monoxide gas is the most dominant pollutants and contributes to the highest Index with 0.23 (SO_2 – CO) followed by 0.18 (for CO–WS) compared to lower interactions were obtained during MCO (2020 data). Higher indexes may also be a good indication of the effects of motor vehicle emissions induced to the atmosphere before MCO. In contrast, SO_2 gases, which usually are emitted from industrial activities or diesel engine operation, is found the topmost dominant influencing the index levels during MCO with 0.17 (SO_2 – O_3), 0.16 (NO-SO_2), 0.15 (NO_x – SO_2) and 0.14 (SO_2 – O_3). The primary *sources* of SO_2 *emissions* are fossil fuel combustion at power plants, refineries, and other industrial facilities (cement, quarrying activities or other industries).

Table 4. Summary of The Strength of Interactions Effects (SIE) between parameters in this study for Pegoh station

	NO - NOx	NOx-SO_2	NO-SO_2	SO_2 -O_3	NO_2 -CO	SO_2 -CO	NO_2 -O_3	SO_2 - Temp	CO-WS
2019	0.10	0.04	0.07	0.05	0.14	0.23	0.08	0.13	0.18
2020	0.01	0.15	0.16	0.17	0.06	0.07	0.13	0.06	0.07
Differ	0.09	-0.11	-0.09	-0.12	0.08	0.16	-0.05	0.07	0.11

Further analyses were done to explore the SIE in a graphical form. The level of concentrations is plotted, and the lined contour displays the same concentration effects within the same concentrations. Figure 2 a-d and 3a-d illustrate the SIE and Index of Interactions between parameters for Pegoh station in 2020. The SIE and Index of Interactions between parameters for Pegoh station in 2019. As described, the highest interactions between CO and SO_2 gases with 0.23 out of 1 indicated that these pollutants usually are emitted from motor vehicles emissions. These results also strongly suggest the potential impact on people's health due to the high level of interactions, especially when finding a strategy to reduce air pollution in an urban environment. It was also indicated that lower interaction had been shown between gases during MCO conditions, which shows there are fewer interactions of gases during MCO time due to lower levels of pollutants.

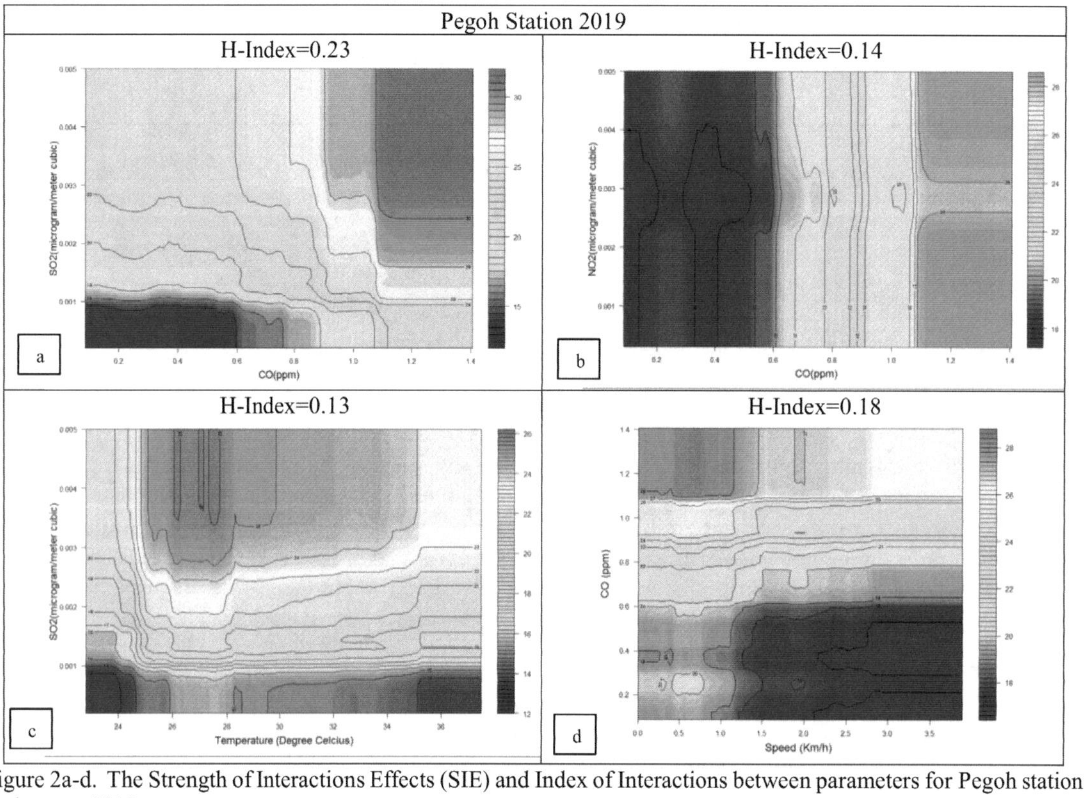

Figure 2a-d. The Strength of Interactions Effects (SIE) and Index of Interactions between parameters for Pegoh station in the year 2019

Pegoh Station 2020

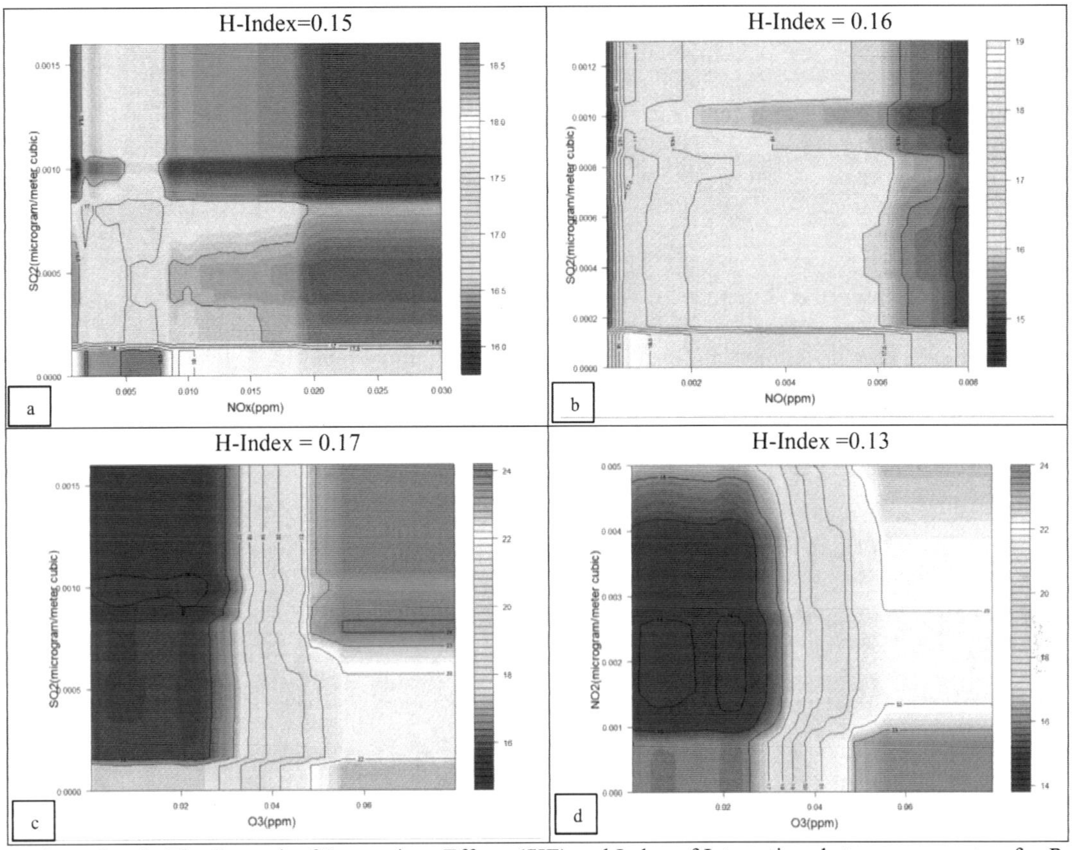

Figure 3 a-d. The Strength of Interactions Effects (SIE) and Index of Interactions between parameters for Pegoh station in the year 2020

4. Conclusions

During the COVID-19 pandemic quarantine, the restricted movement of motor vehicles, transportations, and closure of industries greatly influenced the concentration of air pollutants. Some measures taken during the COVID-19 pandemic period may be turned into permanent habits over time. Comparing the 2019 and 2020 MCO data, it is clear that CO (18.9 %) gas is the most dominant gas followed by SO_2 (14.6 %) of RVI values before during non-MCO period influenced $PM_{2.5}$ concentrations followed by meteorological parameters (relative humidity and wind directions). Similar during MCO, CO remained the highest variable to control $PM_{2.5}$ with 22.6 % RVI, followed by O_3 with 12.1 % and meteorological parameters (temperature and relative humidity). These variables have also been commonly used in other air pollution studies and influence air pollution data, especially during MCO and non-MCO conditions. The highest difference of RVi is shown by SO_2 levels, which reduced from 14.6% during non-MCO to only 5.7 % during MCO periods.

Studying the characteristics of PM_{10}, CO, and O_3 is vital for improving the quality of life, physical health, and inhabitants' living environment. However, the factors that influence these phenomena are still in study in most cities and expanded. The COVID-19 outbreak may be a turning point in combating global climate change as the behaviours of travelling and people's movements are restricted. The improvement in air quality can be considered as the most crucial gain after the pandemic. Air pollution and climate change have long term effects through complex interactions in the atmosphere and ground-level surface of pollutants, especially in urban environments.

It is valuable to understand the relationship between parameters of the RVI, SIE, and IOI to the concentrations of particulate matter. Especially when observing the pattern of before and after the MCO was implemented in certain states or countries. The findings add our knowledge of the mechanism of air pollution and provide a scientific reference for implementing targeted control measures. This study provides an insight into how the gaseous interaction in the atmosphere happens, which might be harmful to human health and the environment in general. For sustainable environmental and air quality, gains from measures should be continued.

Acknowledgments
The authors wish to express their gratitude towards the Department of Environment (Air Division), Ministry of Water and Environment, Malaysia for producing the data, Atmospheric, Air Quality and Climate Change Research Group, University Malaysia Terengganu, University Putra Malaysia, University Kebangsaan Malaysia, Mineral Research Centre, Minerals and Geoscience Department Malaysia, Universiti Sultan Azlan Shah, Malaysia and The Clean Air Forum Society of Malaysia for their support.

References
[1] Guo YR, Cao QD, Hong ZS, et al. The origin, transmission and clinical therapies on coronavirus disease 2019 (COVID-19) outbreak- An update on the status. *Mil Med Res* 2020; 7: 1–10.
[2] WHO. *Coronavirus disease 2019 (COVID-19) Situation Report-94*. 2020.
[3] WHO. Ambient (Outdoor) Air Quality and Health. *World Health Organization*, https://www.who.int/news-room/fact-sheets/detail/ambient-(outdoor)-air-quality-and-health (2018).
[4] Shah AUM, Safri SNA, Thevadas R, et al. COVID-19 outbreak in Malaysia: Actions taken by the Malaysian government. *Int J Infect Dis* 2020; 97: 108–116.
[5] Abdullah S, Mansor AA, Napi NNLM, et al. Air Quality Status during 2020 Malaysia Movement Control Order (MCO) Due To 2019 Novel Coronavirus (2019-nCoV) Pandemic. *Sci Total Environ* 2020; 729: 139022.
[6] DOE Malaysia. New Malaysia Ambient Air Quality Standard, https://www.doe.gov.my/portalv1/en/category/info-umum/info-kualiti-udara (2020).
[7] Latif MT, Dominick D, Hawari NSSL, et al. The concentration of major air pollutants during the movement control order due to the COVID-19 pandemic in the Klang Valley, Malaysia. *Sustain Cities Soc* 2021; 66: 102660.
[8] Marsha A, Larkin NK. A statistical model for predicting PM2.5 for the western United States. *J Air Waste Manag Assoc* 2019; 69: 1215–1229.
[9] Ul-Saufie AZ, Yahaya AS, Ramli NA, et al. Future daily PM10 concentrations prediction by combining regression models and feedforward backpropagation models with principle component analysis (PCA). *Atmos Environ* 2013; 77: 621–630.
[10] Carslaw DC, Taylor PJ. Analysis of air pollution data at a mixed source location using boosted regression trees. *Atmos Environ* 2009; 43: 3563–3570.
[11] Yahaya NZ. *Temporal and Spatial Variations of Ultra-fine Particles in the Urban Environment*. University of Leeds, 2013.
[12] Yahaya NZ, Ghazali NA, Ahmad S, et al. Analysis of daytime and nighttime ground level ozone concentrations using boosted regression tree technique. *Environ Asia* 2017; 10: 118–129.
[13] Yahaya NZ, Ibrahim ZF, Yahaya J. The used of the Boosted Regression Tree Optimization Technique to Analyse an Air Pollution data. *Int J Recent Technol Eng* 2019; 8: 1565–1575.
[14] Friedman JH. Greedy Function Approximation: A Gradient Boosting Machine. *Ann Stat* 2001; 29: 1189–1232.
[15] Friedman JH. Stochastic gradient boosting. *Comput Stat Data Anal* 2002; 38: 367–378.
[16] Munir S, Luo Z, Dixon T. Comparing different approaches for assessing the impact of COVID-19 lockdown on urban air quality in Reading, UK. *Atmos Res*; 261. Epub ahead of print 15 October

2021. DOI: 10.1016/J.ATMOSRES.2021.105730.

[17] R Core Team. R: a language and environment for statistical computing. *R Foundation for Statistical Computing*, https://www.gbif.org/tool/81287/r-a-language-and-environment-for-statistical-computing (2018).

[18] David Carslaw. *The openair manual open-source tools for analysing air pollution data*, https://github.com/davidcarslaw/openair. (2019).

[19] Carslaw DC, Ropkins K. openair — An R package for air quality data analysis. *Environ Model Softw* 2012; 27–28: 52–61.

[20] Ridgeway G. Generalized Boosted Models: A Guide to The GBM Package, https://rdrr.io/cran/gbm/man/gbm.html (2007).

[21] Ridgeway G. Generalized Boosted Models: A Guide to the GBM Package, https://www.researchgate.net/publication/215990688_Generalized_Boosted_Models_A_Guide_to_the_GBM_Package (2005).

[22] De'ath G. Boosted Trees for Ecological Modeling and Prediction. *Ecology* 2007; 88: 243–251.

[23] Akima H. Algorithm 761: Scattered-data surface fitting that has the accuracy of a cubic polynomial. *ACM Trans Math Softw* 1996; 22: 362–371.

[24] Akima H. A Method of Bivariate Interpolation and Smooth Surface Fitting for Irregularly Distributed Data Points. *ACM Trans Math Softw* 1978; 4: 148–159.

A review of plastic and microplastic pollution towards the Malaysian marine environment

E Tan[1], N F Jaafar[2], S H Aileen Tan[1] and N B Mohd Zanuri[1*]

[1]Centre for Marine and Coastal Studies (CEMACS), Universiti Sains Malaysia, 11800 Gelugor Penang.
[2]School of Chemical Sciences, Universiti Sains Malaysia, 11800 Gelugor Penang.

*Corresponding Author email: lailazanuri@usm.my

Abstract Plastics are widely used in daily lives but uncontrollably dumped into the ocean by humans. Plastic pollution is harmful to the marine environment and organisms as it can break down into microplastics (MPs) and release chemicals into the water. Microplastics (MPs) are small, fragmented plastic pieces (< 5mm) that exist in every part of the ocean. MPs are problematic because they are hard to recover and can be easily consumed by marine organisms, resulting in bioaccumulation and biomagnification in the food chain. Malaysia is a tropical country located at the heart of Southeast Asia (SEA) and owns diverse marine ecosystems and organisms. Malaysia is affected by plastic pollution due to rapid development and intense economic activities. In this paper, we discuss the plastic pollution crisis in Malaysia and its contributing factor. We review the possible effect on the marine environment in Malaysia. Knowledge gaps to manage plastic pollution in Malaysia are also addressed. As MPs are consistently discovered in different compartments of the Malaysian marine environment, there is an urgency to develop a better waste management system and strong cooperation from all societal levels to handle plastic pollution in the country.

Keywords: Plastic, Microplastics (MPs), Tropical, Malaysia, Southeast Asia (SEA)

1. Introduction

Plastics are artificial products made of synthetic polymers from petroleum [1]. They are omnipresent in today's ocean; from the surface to the bottom ocean, occurring in different sizes [2]. It has been estimated that approximately 8 million plastic enters the ocean annually [3]. The buoyancy of plastics alleviates its impact as a pollutant because it can be easily windblown from land to the ocean and carried by the surface current [4]. When exposed to UV radiation, plastic undergoes oxidation, causing them to become brittle and broken down into smaller pieces [5]. These small plastic fragments are generally known as microplastics (MPs), characterized by <5 mm in diameter and can be categorized into small MPs (<1 mm) and large MPs (1-5 mm) [6]. Plastic was initially discovered in the ocean around the early 1970s but it did not steal the world's spotlight as a marine pollutant until the term 'microplastic' was introduced by Thompson and his colleagues in 2004 [7]. Since then, reports on its origin, distribution and fate have been gradually studied [6]. The exposure of aquatic animals and plants towards plastics are life-threatening as it can physically entangle marine organism, preventing them from swimming and causing suffocation, leading to death [8]. Hence, the objectives of this paper are to summarize the plastic pollution crisis in Malaysia, discuss the possible effect of plastic and MPs pollution on the marine environment in Malaysia and identify knowledge gaps in managing plastic pollution in Malaysia.

2. The Malaysian Waters

Malaysia is strategically located at the heart of Southeast Asia (SEA) (Figure 1) and encompassed by large water bodies with only the northern tip of Peninsular Malaysia connected to the Asian landmass via Thailand (Figure 2) [9]. Although the country is made up of only 0.2% of the world's landmasses, it is ranked 12th in the list of the world's 17 megadiverse countries by the National Biodiversity Index [10,11]. The Malaysian water coverage is estimated at around 453,186 km^2 [12]. Sitting at the equator, the Malaysian marine environment is characterized by warm waters and sunny conditions all year round with two monsoon seasons: Northeast and Southwest monsoon [13]. In efforts to protect the marine environment, the government gazetted 50 locations around the country (42 in Peninsular, five in Sabah and three in Sarawak) as Marine Protected Areas (MPAs) which covers approximate 1.4% of the Malaysian waters [12]. As a developing maritime country and owning one of the busiest international shipping routes, Malaysia is highly dependent on ocean-based economic activities (e.g., shipping and tourism) to support the country's economic growth [14]. In addition to staggering terrestrial development and industrial activities occurring in the country, the marine environment is experiencing enormous pressure and severe destruction like never before and this includes population density increment, lack of environmental awareness among the public, tourism activities, coastal conversion for aquaculture activities and land reclamation [12].

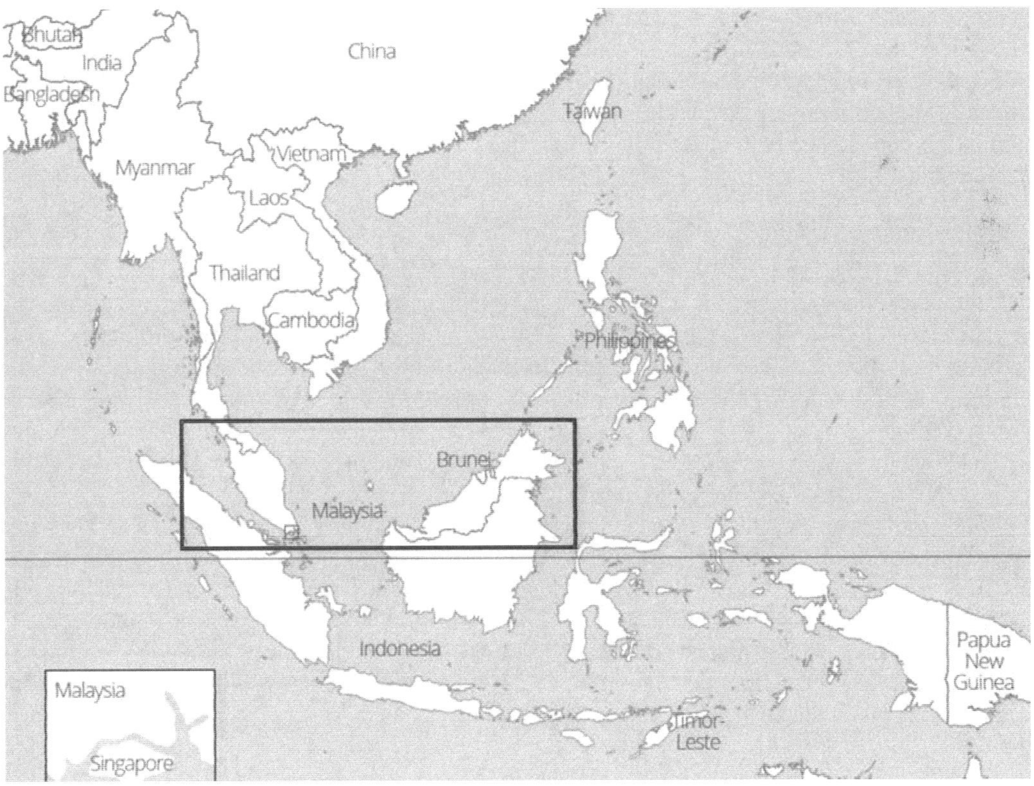

Figure 1. Southeast Asia (SEA) map [15]. Malaysia is located at the heart of the SEA region (highlighted in the box).

Figure 2. Detailed map of Malaysia [16]. (A) The northern part of Peninsular Malaysia connected to Thailand. (B) Malaysia sitting closely at the equator line.

3. Plastic problem in Malaysia

The Asia continent is the largest plastic manufacturer and waste importer [17]. Upon the implementation of the National Sword Policy in 2018, China as the largest plastic waste import in the world has caused a huge industrial shift to Southeast Asia (SEA) countries as the world's new dumping site, including Malaysia [18]. Furthermore, developing countries including Malaysia often encounter plastic waste accumulation because of high processing costs and advanced technology additional which do not guarantee high economic returns as recycled plastics are not well-received by industries and consumers because of their poor quality [19]. Statistics have shown that where Malaysia was ranked 8th globally as top plastic polluters in 2021 and holds the record of Asia's highest plastic consumption at a rate of 62kg/capita/year [20]. The country is home to more than 1300 plastic manufacturing factories, owning one of the largest plastic industries in the world [24] with an average rapid growth rate of 15% within 11 years [21]. As local authorities realized the environmental problems created due to the exponential growth, at least 100 illegal factories were forcefully shut down [22]. At the same time, strict import regulation was enforced by authorities when plastic import to the country skyrocketed in 2018 to 870 000 tons, resulting in lower import numbers at 143 000 tons [22]. The marine environment in Malaysia is extremely vulnerable to pollution with rapid coastal development, economic and maritime activities, especially the coastline of Peninsular Malaysia [23].

In Malaysia, plastic waste is managed by the National Solid Waste Management Department (NSWMD) under the Ministry of Housing and Local Government Housing (MHLG) [24]. The current method in tackling plastic waste is either by domestic burning, landfills disposal or recycling [24]. Recycling efforts have been emphasized in the National Solid Waste government's policies and the target was set at 22% in 2020 and has reached 30% in the same year and aims to hit 40% by 2025 [25]. Figure 3 summarized the factors that cause the country's plastic problem while the following segment describes the factors in detail:

3.1 Weak Management System and Law Enforcement

The main driver to the severity of plastic pollution in Malaysia is the flawed waste management and recycling system in the country [26]. The country often relies on landfills to dispose waste which is extremely inefficient towards persistent plastics [26]. Besides being dumpsites to developed countries such as Australia, Canada, UK and US, the Malaysian recycling factories is incompetent in recycling huge amount of plastic because they operate illegally without adequate technology and inadequate waste sorting at the societal level reduces recycling effectiveness [27].

3.2 Conflict in Political Views

In 2018, the Energy, Technology, Science, Climate Change and Environment (MESTECC) ministry of Malaysia was concerned about the environmental degradation of plastic waste but the minister from the Ministry of Housing and Local Government (MHLG) expressed that the plastic import and recycling industry can potentially generate high economic benefits to the country [28]. Thus, both ministries together with the Ministry of International Trade and Industry and Industry (MITI) and Water, Land and Natural Resources (KeTSA) have worked out to implement strict enforcement on factories to import only clean and high-quality plastics and shut down plastic processing factories that do not comply to the government's requirement [29]. In addition to inspection from local authorities, 139 factories operating illegally have been shut down as of February 2019 [30] with only 62 Malaysian companies holding approved permits (AP) to import and process plastic waste into the country [31].

3.3 Low Awareness in Society

Emerging consumerism and capitalism with a lack of knowledge about plastic pollution among Malaysian increases consumption of SUPs and other plastic products in daily lives [32]. At the same time, environmentally sustainable products available in the country are not well-promoted among consumers and expensive compared to conventional ones [22].

3.4 Limited Recycling Capabilities

Out of seven classes of plastics, Malaysia is only capable of recycling three classes (Type 1: PETE; Type 2: HDPE; Type 4: LDPE) [18]. On top of that, the low recycling rate is unable to keep up with the consumption rate of consumers further causing more plastic waste accumulation in the country [22]. In addition, common waste such as plastic products, polystyrene, diapers and feminine towels are directly thrown into landfills, rather than properly recycled [33].

3.5 Lack of Uniformity in Local Supervision

Some recreational beaches in the country are taken care of by external contractors hired by the state or local municipal authorities themselves [23]. Without standardized administration by the federal government across the country, the difference in control and management may lead to some marine sensitive areas being mishandled or overlooked by non-professionals.

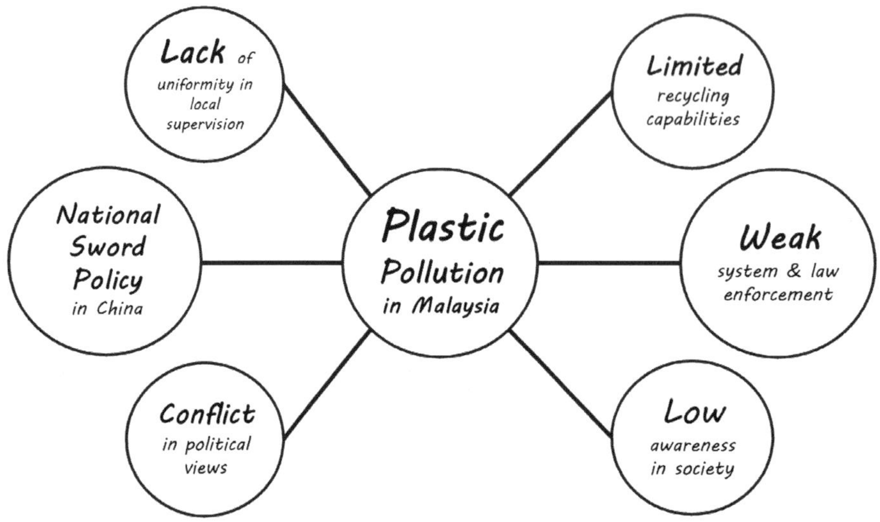

Figure 3. Plastic pollution in Malaysia. Six dominating factors are identified as contributing factors to the country's plastic pollution.

4. Plastic pollution towards the marine environment in Malaysia

As part of the Coral Triangle marine region, the 4800 km coastline of Malaysia is home to numerous highly productive marine ecosystems such as coral reefs, mangrove forests, mudflats, seagrass meadows and white sandy beaches that harbors valuable marine organisms such as dolphins, dugongs, marine birds and sea turtles [34]. Nevertheless, these fragile marine resources are susceptible to habitat destruction and environmental pollution due to weak local law enforcement, inadequate plastic waste management, intense coastal development and terrestrial activities [35]. In Malaysia, the responsible authorities managing marine resources includes the Department of Fisheries and Department of Marine Park Malaysia (Peninsular Malaysia); the Sarawak Forestry (Sarawak); and the Sabah Parks and Sabah Wildlife Department (Sabah) [36]. Figure 4 outlines the marine mammals and ecosystems discussed in this paper about plastic pollution.

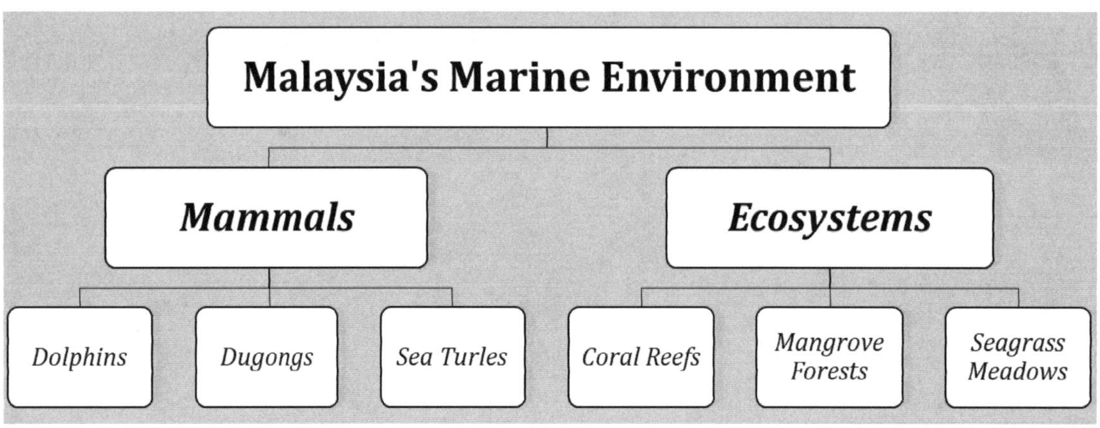

Figure 4. Important components of the Malaysian marine environment. The impact of plastic and MPs pollution on these marine mammals and ecosystems in Malaysia is discussed in this paper.

4.1 Marine Mammals in Malaysia

Marine mammals being long-lived and dominant in the higher level in food webs serve as a reliable bioindicator because pollutants can accumulate in the body [37]. When mammals ingest plastics, it blocks important pathways for food transfer and air exchange whereas entanglement impedes their movement in the water, causing drowning and ultimately death [8]. Meanwhile, MPs can enter marine mammals through several pathways: consumption of contaminated prey, direct uptake through filter-feeding or inhalation at the sea surface [38,39]. Although the entry of MPs into the body would unlikely to result in serious suffocation or threaten their survival relative to their body size, persistent and toxic chemicals present in plastic nevertheless would eventually bring damage at a cellular level, reduce their physiological function and eventually population level [40]. [37] reported that marine mammals that died because of infectious diseases have higher MPs concentration in the body, compared to others that died due to other natural factors, suggesting that MPs may have carried pollutants and caused infection internally.

Marine mammals in Malaysia in particular, are highly exposed to MPs via living in coastal water bodies and feeding on planktons, seagrasses and benthic invertebrates such as sea cucumbers and bivalves, where MP is also highly available in their habitats and preys [41]. Transfer of MPs and associated chemicals would undoubtedly occur when these marine mammals unknowingly consume contaminated prey or directly from water and food sources [38]. This would also affect their development and reproduction as toxic chemicals bioaccumulate in their body [38].

There are news reports on the death of marine mammals due to plastic debris consumption or entanglement in the country. Often, when the carcasses ended up on shore, they are buried immediately after inspection by government authorities or sometimes send to research centers for an autopsy if research facilities are available nearby [42]. According to autopsies carried out by the Malaysian Marine Fisheries Mammals Network's National Task Force, a significant amount of plastic waste can be found in marine mammals and plastic pollution has become the emerging contributor of mortality in them [43]. The following passage highlights several notable marine organisms in Malaysian waters and the plastic impact on them.

4.1.1 Dolphins. There are three confirmed dolphin species found in Malaysian waters: Indo-Pacific finless porpoise (*Neophoceana phocaenoides*), Indo-Pacific humpback dolphin (*Sousa chinensis*) and Irrawaddy dolphin (*Orcaella brevirostris*) [44]. Their preferred living condition is shallow coastal waters, about 3-25 m deep such as mangroves, estuaries and swamps [45,46]. Despite the lack of published data on dolphin species in Malaysia, news reports and pictures from social media prove the existence of a diverse range of dolphins in the country and its surrounding waters. This includes reports of Indo-Pacific bottlenose dolphin (*Tursiops aduncus*) stranded and released back safely into Kuala Perlis waters [46], a carcass of Roughed-tooth dolphin (*Steno bredanensis*) washed up to the coast of Kota Belud, Sabah [47], Pantropical-spotted dolphin (*Stenella attenuata*) observed in the waters of Sarawak during an expedition [48], injured Spinner dolphin (*Stenella longirostris*) being washed up the shore and treated in Tuaran, Sabah [49] and Stripped dolphins (*Stenella coeruloealba*) occurring in the northern Malacca Straits [50]. The Indo-Pacific bottlenose dolphins and spinner dolphins were also occasionally seen in the waters of Tun Mustapha Marine Park (TMP) in Sabah [51]

Dolphins in Malaysian waters are primarily threatened by coastal development, destructive fishing practices such as trawling and fish bombing (still widely practiced in Sabah), underwater noise pollution due to high vessel traffic and the mishandling of plastic waste that causes habitat pollution in nearby coastal water bodies [51]. For instance, the waters of Penang Island are crucial breeding grounds for Indo-Pacific finless porpoise as mothers and young Irrawaddy dolphins can be seen in the area [52]. However, the island is undergoing land reclamation, rapid coastal development involving the conversion of mangrove estuaries that are increasingly threatening their habitat [52,53]

Plastic wastes are also gradually becoming the prominent culprit that causes death in dolphins. In 2015, a dolphin carcass from Likas Bay, Sabah underwent autopsy and 4.25kg of plastic waste was recovered from the body. Researchers suspected that the dolphin died from starvation as the plastics have disrupted the pathway for food transfer [54]. In later years, a dolphin carcass was found on a remote beach in Penang Island with death attributed to plastic entanglement [42]. Dolphin death due to plastic pollution is expected to increase in the future with a high amount of plastic waste entering the Malaysian water bodies and no efficacious action taken by the government to resolve the plastic problem in the country.

4.1.2 Dugongs. Dugongs (*Dugong dugon*) are herbivore marine mammals, primarily feeding on seagrasses, usually live-in shallow tropics and subtropics coastal waters (East Africa to Indo Pacific region) [55]. Also known as sea cows, they occasionally consume algae and invertebrates [56]. In Malaysia, the dugong population is protected by law under the legislation of the Department of Fisheries Malaysia: Fisheries Act 1985 Part IV and the Fisheries (Control of Endangered Species of Fish) Regulations 1999 since 2011 [56]. Dugongs are exclusively found in three different states in Malaysia: Johor (Pulau Sibu and Pulau Tinggi), Sabah (Pulau Bangi, Pulau Mantanani and Pulau Mengalum) and Sarawak (Brunei Bay) [56,57].

Dugongs usually occur in coastal regions where seagrass meadows exist as they are their key diet, breeding and nursing ground. They are often observed in Lawas, Sarawak with the existence of high diversity seagrass species in the waters [57]. Dugongs have extremely low reproductive rates and require high dedication to nurture their calves [58]. Researchers have pointed out that the dugong population increases merely at a maximum rate of 3% annually only if they live in conditions with minimal anthropogenic and are not subjected to human-caused mortality [55].

A growing amount of plastic waste in Malaysian waters is anticipated to be one of the contributing factors to dugongs' death [56]. In Malaysia, there is so far one news report of a male dugong who died due to fishnet entanglement at Pulau Sibu, Johor [59]. In Thailand, an 8-month baby dugong 'Marium' was found dead due to ingestion of plastic waste which causes gastritis and blood infection [60].

The capability of seagrass meadows to accumulate MP will ultimately threaten the survival of the dugong population as chemicals released from MPs can gradually accumulate in the body [56]. Dugongs are currently classified as Vulnerable on the IUCN Red List and without stricter effort to protect seagrass meadows and refine their living conditions, dugongs will eventually disappear in the Malaysian waters [56,61].

4.1.3 Sea turtles. Malaysia is one of the critical nesting grounds in the world for sea turtles where four out of seven marine turtle species land here [62]. This includes the green turtles (*Chelonia mydas*), hawksbill turtles (*Eretmochelys imbricata*), leatherback turtles (*Demochelys coriacea*) (currently no more landings) and Olive Ridley turtles (*Lepidochelys olivacea*) [62]. In Peninsular Malaysia, sea turtles can be found in the states of Terengganu, Melaka and Johor with Terengganu having the highest nesting rate of at least 2000 per year [56,62]. Meanwhile, in East Malaysia, the Sabah's Turtle Park at the Borneo Island welcomes at least a total of 10000 sea turtle nests of green and hawksbill turtles annually, making the state the leading nesting ground in the SEA region [62].

The aftermath of floating plastic debris in the ocean towards sea turtles is so profound to the extent that all seven species around the world have been reported to be involved in plastic entanglement while swimming or ingestion due to its high resemblance to jellyfish [8]. The major threat that sea turtles face in Malaysia is the illegal trading of turtle eggs for domestic and international demand and the brutal killing of sea turtles for food and medicinal purposes [62-66]. In recent years though news reports on sea turtles drowning due to plastic ingestion and entanglement of discarded fishing gear are becoming more common in the country [67]. The first appearance of an entangled Loggerhead turtle (Caretta caretta) in Malaysian waters by fishing nets was also documented recently [68].

Additionally, MP accumulation on sea turtle's nesting beaches threatens their survival because the incubation temperature of turtle eggs relies on the insulation ability of sandy beaches [38,70]. The sand temperature was reported to decrease by 0.75°C when concentration MP is 1.5 to sand, influencing the sex of the hatchlings [38]. When MPs are mixed with sands, the sediment permeability increases and a longer time is required to heat up and reach the optimal temperature for hatchlings incubation [38]. The disappearance of the leatherback turtles in Malaysia was a real-life example of failure to achieve optimal incubation temperature, influencing the sex ratio of hatchlings with a higher proportion of female individuals, leading to population shift [69]. Although this tragedy was solely caused by human error (limited knowledge in the early days), nevertheless it highlighted the importance of reaching optimal incubation temperature in the sand so that the sex ratio balance is not interrupted. Hence, it is essential to ensure sandy beaches and shores are kept as clean as possible away from plastic debris and MPs [38]. In Malaysia, [70] has reported the occurrence and recovered a total of 2489 pieces of MPs in four sandy beaches in Terengganu which are all important nesting areas for sea turtles and emphasized that it is the importance of further studies on MPs contamination on nesting and landing beaches for sea turtle.

4.2 Marine Ecosystems in Malaysia

Coral reefs, mangrove forests and seagrasses meadows are the dominating ecosystems occurring along the Malaysian coastline [20]. Conservation efforts to protect these ecosystems should be simultaneous and inclusive as they often coexist in the marine realms and have a strong connection in terms of ecological functions as summarized in Figure 5 [71]. A study revealed that muddy sediments of sheltered seagrass meadows with coral reefs were found to have higher carbon content and organic material compared to exposed meadows without coral reefs [72]. Coral reefs were also reported to act as a barrier to enhance seagrass meadows capacity as a blue carbon sink locally to accumulate carbon materials in the sediment by slowing down water current and reducing sediment erosion rate [72]. Furthermore, they are also critical impact absorbers during huge storms, waves or natural phenomena such as tsunamis which help protect coastal communities [71]. The following paragraphs explore the impact of plastic pollution on three ecosystems present in the tropical waters of Malaysia.

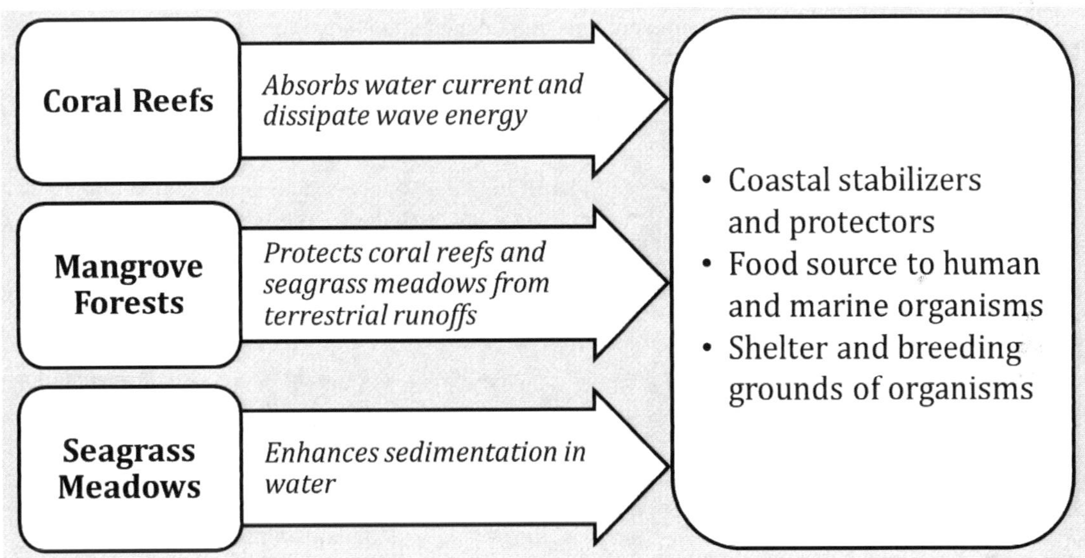

Figure 5. The importance of the coastal ecosystem found in Malaysia**.** Respective and shared ecological functions of coral reefs, seagrass meadows and mangrove forests towards humans and the environment.

4.2.1 Coral Reefs. Coral reefs are well known to be highly productive, providing food and breeding grounds for numerous marine invertebrates and vertebrates [73]. With intense coastal development around the world and imprecise law enforcement to protect the coral reefs distributed along the coast or near land, they are prone to severe destruction and pollution [71].

The Malaysian waters have exceptional coral species with a total estimation of a minimum of 360 species in West Malaysia and 500 species in East Malaysia, excluding deep-sea reefs and atolls within the continental shelf in Malaysia in the South China Sea (e.g., Spratly Island and Luconia Shoals) [35]. The coral reefs in Sabah are considered to be well protected under law as vicinity waters are gazetted as marine parks for conservation [35]. Over in Peninsular Malaysia, the coral reefs at the west coast in the

Straits of Malacca are less healthy compared to the east coast facing the South China Sea because of busier maritime activities [35].

Non-governmental organization Reef Check Malaysia is mainly responsible for the statistical distribution and health status analysis of coral reefs in Malaysia [74,75]. The coverage of living corals was at 42.42% in 2018, decreasing to 40.63% in 2019 with a slight increase to 41.32% in 2020 [74,75]. The health of coral reefs in Malaysia was reported as 'Fair' and they are mainly influenced by rapid human activities, coastal developments, terrestrial input of waste and weak waste management of the country [74,75]. A baseline study focusing on the distribution and impact of marine debris towards coral reefs in Darvel Bay, Sabah revealed the density of 10 plastic items within $100m^2$ perimeter and the authors highlighted the importance of quantifying the abundance of macro-plastic debris in future work (monitoring programs and research) as plastic waste may deteriorate the health and well-being of coral reefs [20].

Coral reefs can also be affected by the MPs in the water column because they practice filter-feeding, relying on tentacles to draw food to themselves and MPs can be accidentally captured and consumed [76]. Certain species of hard corals (Scleractinia) in Australia were reported to ingest MPs particles and are capable of consuming as much as 50 µg of plastics [77]. It has also been suggested that MPs can damage the corals internally when consumed despite being observed to expel within 48 hours [78]. Nevertheless, the negative effect from MPs consumption towards corals remained uncertain [77].

Malaysian corals located in coastal areas are prone to sedimentation, tourism activities, shipping industries and coastal development which would increase MPs concentration into coastal waters [79]. The above findings strongly suggest the negative impact on plastic pollution towards coral reefs requires stern pollution control and management by local authorities to secure optimal water quality conditions for survival.

4.2.2 Mangrove forests. Mangrove forests are exclusively distributed in shallow tropical and subtropical waters, coexisting with coral reefs [71]. Often found in salty intertidal zones, mangrove trees are unique because of their tolerance towards coastal tide fluctuation, high water salinity and the ability to stay submerged in waters for a long time [72]. Highly productive mangrove forests also provide food, shelter and breeding ground for fishes, birds and invertebrates (e.g., crustaceans, molluscs) [73]. They also protect surrounding coastal communities by absorbing the impact of strong wind and waves during storms [74].

Although SEA owns the highest abundance of mangrove species globally, mangrove forests in this region experiences a high destruction rate at about 30% loss since 1980 [74]. Malaysia is ranked 3rd globally for mangrove forest coverage, with 85% of 570 000 hectares being gazetted and protected under Malaysian law [34]. The estimated distribution of mangrove forests in Peninsular Malaysia, Sarawak and Sabah are 18%, 22% and 60% respectively [34,75]. Mangrove forests are famously used as the country's ecotourism spot to watch fireflies at night in Matang Mangrove Forest Reserve (Perak), Kampung Kuantan Firefly Park (Selangor) Sungai Kerian (Pulau Pinang) [76].

Estimated to own at least two-thirds of the world's mangrove species [34], mangrove forests in Malaysia also support a wide variety of marine mammals, birds and terrestrial organisms such as the fireflies, the dusky leaf monkey and the Proboscis Monkey [71]. As one of the stopover locations for migratory birds along the East Asian-Australian Flyway, mudflats near mangrove forests are temporary habitats and breeding grounds for migratory marine birds as they avoid the harsh winter weather and predation [77]. Several important marine bird species that can be found throughout the Malaysian mangrove forest include Near Threatened Grey-tailed tattler (*Tringa brevipes*), Endangered Nordmann's Greenshank (*Tringa guttifer*) and Near Threatened Oriental Darter (*Anhinga melanogaster*) [73, 78-80].

The major threats to Malaysian mangroves include conversion and clearcutting of mangroves trees for agriculture or aquaculture activities, timber manufacturing industry and careless terrestrial waste disposal [81]. Mangrove forests in Malaysia and marine organisms living in this ecosystem are highly susceptible to plastic pollution because they are located at the intersection of land and ocean, where they can receive a considerable amount of waste from factories and wastewater treatment plants built nearby rivers [71]. Plastic waste will pass through the waters of mangrove forests before entering the ocean [71]. Marine birds being reported to ingest MP at coastal waters because of their opportunistic feeding nature will also be threatened as migratory birds stopping over in Malaysia [82]. In recent years, MP research in the country is thriving and work has been done on several compartments in mangrove estuaries such as organisms [83-87], sediments [88-90] and waters [90,91].

4.2.3 Seagrass meadows. Seagrasses are submerged halophytes found in enclosed coastal regions that reproduce vegetatively to form seagrass meadows [34]. It has been estimated that the seagrass destruction rate stands at 45% per year throughout SEA due to extensive coastal development and water pollution [92]. Similarly in Malaysia, seagrass was previously reported to be extensive and abundant along its Peninsular coastline but current statistics have shown significant loss [34]. Also ranking 3rd globally in terms of diversity, 16 seagrass species are present in the Malaysian waters: *Cymodocea rotuntada*, *Cymodocea serrulata*, *Enhalus acoroides*, *Halodule pinifolia*, *Halodule univernis*, *Halophila beccarii*, *Halophila decipiens*, *Halophila major*, *Halophila minor*, *Halophila ovalis*, *Halophila sp.*, *Halophila spinulosa*, *Ruppia maritima*, *Syringodium isoetifolium*, *Thalassadendron ciliatum*) [34,35] with Halodule and Halophila species as dugong's favourite diet [55].

At present, seagrass meadows in Malaysia can be found scattered in the states of Johor, Sabah and Sarawak coastal waters [56]. Seagrasses are exceptionally important as they are the main diet and habitat to vulnerable marine mammals such as dugongs and sea turtles [93]. Besides hosting a wide range of marine organisms, seagrasses meadows are also bioengineers where they redistribute and dissipate energy from strong ocean waves and storms [94]. Recently, seagrasses are discovered to elevate the effect of climate change by sequestering approximately 10% of atmospheric carbon dioxide into the marine environment [92].

As seagrasses meadows occur in shallow waters adjacent to land, they are subjected to high input of terrestrial waste and could potentially act as MP sinks because they can trap particles easily [95]. In the same report, the accumulation rate of MP in seagrass meadows sediments was observed to enhance at least twice faster in vegetated sediments compared to non-vegetative ones. If seagrass meadows are located in semi-enclosed areas, sediment beds are even more prone to MP accumulation as seagrass leaves can reduce particle momentum and cause them to sink to the bottom [95]. Plastic density also affects the trapping effect at seagrass meadows. A study found out that low-density MPs (PE, PP and PVC) ranging from 0.55 – 287.00 mm in seagrasses (*Posidonia oceanica*) are capable of forming aggregates by using lignocellulosic fibers and carried to shore by wave action while high-density MPs will instantly sink to seagrass beds [96]. Additionally, [97] reported that lighter MP is less likely to be trapped in seagrass meadows compared to heavier ones because of the lower sinking rate in the water column, which allows it to not be easily trapped to seagrass blades by using biofilms. Gradual MP accumulation at seagrass meadows has also been reported to change sediment composition which may affect seagrass growth [95] and threaten the health of marine organisms living near the meadows with increasing MP exposures [94].

Seagrasses as native plants in the Malaysian waters are unfortunately not fully protected by law; only those in waters of gazette marine park areas are protected [56]. With decreasing dugongs and vulnerability state of seagrass meadows in Malaysia, comprehensive preservation efforts are crucial to protect these delicate plants, which would also indirectly safeguard the wellbeing of marine organisms living within them [56]. More research effort is needed to study the condition and delineate the pollution status of seagrasses, particularly to MPs as plastic waste is likely to increase exponentially and indirectly pressures the cleanliness of nearby coastal waters and the wellbeing of marine organisms [34].

5. Collective efforts and the way forward: Malaysia Roadmap to Single Use Plastic 2018-2030

The journey of Malaysia towards reducing plastic reliance initially started in Penang state by introducing a ban every Monday in 2009, followed by Selangor in 2010 on every Saturday. In 2011, the 'The No Plastic Bag Day' campaign was launched and implemented across the country every Saturday with plastic bags upon request charged at RM 0.20 (approximate USD 0.05) [98]. However, the campaign has seemed lost its effectiveness as the charge imposed is not heavy enough to stop the public from using plastic bags during shopping and Single Used Plastics (SUPs) usage is sometimes unavoidable in eateries for hygiene purposes [99]. In 2021, Penang state step up by charging RM 1 per SUP bag at large departmental and grocery stores [100].

In 2018, the newly formed MESTECC developed a comprehensive policy 'Malaysia Roadmap to Single Use Plastic 2018-2030' to specifically address SUPs issue in the country. This is the very first initiative from the Malaysian government that involves all societal levels from individual consumers to government officials to play their part towards environmental sustainability. This policy paves the way to completely remove SUPs reliance on society and introduce eco-friendly solutions and packaging alternatives to the Malaysian community. The ultimate focus of this roadmap is to entirely abolish the use of SUPs in Malaysia by the end of 2030 [32]. The government also started introducing the 'Straw Ban' campaign in Selangor and Putrajaya in 2019 and continuously throughout Malaysia in 2020 [98]. This campaign unfortunately also went into vain as it was implemented on merchants rather than consumers and the lack of consistent enforcement and monitoring by local authorities has resulted in the reappearance of a plastic straw in eateries currently [101].

6. Conclusion and Recommendations

Increased plastic waste invading the marine environment is placing these exquisite mammals and fragile ecosystems in jeopardy. Growing evidence on the repercussions of plastic debris and its associated chemical to different marine compartments across various trophic levels serves as a reminder that plastic pollution will threaten the survival of marine life and would eventually affect the well-being of human health. Coral reefs, mangrove forests and seagrass meadows are the dominant ecosystems along the coastline are impeccably valuable from the environment or human perspective. Conserving and protecting them should be prioritized in the country because once they are lost, the marine environment would collapse as organisms will lose their habitat and food source. Coastal communities might even completely lose their livelihood and protection from natural disasters.

It is anticipated that plastic waste would eventually be the prime reason for the death of marine mammals in the near future. In consideration of their shy personality, low breeding rate and population in natural waters, it is important to carry out more study that identifies their hotspot for breeding, foraging and shelter and enforce strict measurements around the area to maximize protection from plastic pollution. Consistent monitoring within these areas should be implemented to ensure emergency operation or conservation efforts can be executed in the quickest time to minimize harmfully and prevent irreversible effects. Working closely with researchers, allocating sufficient funds and facilities for conservation-related research should also be emphasized by the government to build up a database for future research and to narrow knowledge gaps about the marine environment, particularly Malaysia.

On a final note, our paper hopes to bring awareness to policymakers of the deteriorating condition of Malaysia's marine environment and look forward to more support from the government and the development of more effective conservation efforts to elevate plastic pollution. Although terrestrial development and generation of domestic waste are unavoidable, what we can do is to find a balance between both sides where development can occur at an acceptable and minimizing the waste entry into the ocean.

Acknowledgments

This study was fully supported by the Bridging Incentive Grant project through project code 304/PPANTAI/6316350 code BI0043/2019/0118 and the Ministry of Higher Education Malaysia for Fundamental Research Grant Scheme with project code: FRGS/1/2020/0770/F01786 and Universiti Sains Malaysia. We would like to thank the staff of the Centre for Marine and Coastal Studies (CEMACS) and the School of Distance Education, Universiti Sains Malaysia for their help and guidance when we were conducting this research.

References

[1] Kumar S, Rajesh M, Rajesh K M, Suyani, N K, Ahmed Rasheeq A and Pratiksha K S 2020 Impact of Microplastics on Aquatic Organisms and Human Health: A Review *Int. J. Environ. Sci Nat. Resour.* **26** 59-64
[2] Kibria G 2017 *Plastic Waste, Plastic Pollution – A Threat to All Nations* Project: Pollution and Climate Change Impacts
[3] Galloway, T S and Lewis, C N 2016 Marine microplastics spell big problems for future generations. *Proc. Natl. Acad. Sci. U. S. A.* **113** 2331-33
[4] Lebreton L, Slat B, Ferrari F, Sainte-Rose B, Aitken J, Marthouse R, Hajbane S, Cunsolo S, Schwarz A, Levivier A, Noble K, Debeljak P, Maral H, Schoeneich-Argent R, Brambini R and Reisser J 2018. Evidence that the Great Pacific Garbage Patch is rapidly accumulating plastic *Sci Rep.* **8** 4666
[5] Akdogen Z and Guven B 2019 Microplastics in the environment: A critical review of the current understanding and identification of future research needs *Environ. Pollut.* **254** 113011
[6] Zhang S, Wang J, Liu X, Qu F, Wang X, Wang X, Li Y and Sun Y 2019 Microplastics in the environment: A review of analytical methods, distribution, and biological effects. *Trends Analyt Chem* **111** 62-72
[7] Li P, Wang X, Su M, Zou X, Duan L and Zhang H 2020 Characteristics of Plastic Pollution in the Environment: A Review. *Bull Environ Contam Toxicol* **107** 577-84
[8] Kühn S, Rebolledo E L B and van Franeker J A 2015 Deleterious Effects of Litter on Marine Life *(Marine Anthropogenic Litter)* M Bergmann et al (Bremerhaven: Springer Open) chapter 4 pp 75-116
[9] Mazlan A G, Zaidi C C, Wan-Lotfi, W M and Othman B H R 2005 On the current status of coastal marine biodiversity in Malaysia *Indian J. Mar. Sci.* **34** 76-87
[10] Ministry of Natural Resources and Environment (NRE) 2006 *Biodiversity in Malaysia* p 32
[11] Tong P S 2020 More policies and laws, is it better for biodiversity conservation in Malaysia? *Conservation Science and Practice* **2** e235
[12] Ministry of Natural Resources and Environment (NRE) 2016 *National Policy on Biological Diversity 2016-2025* p 116
[13] World Bank Group (WB) and Asian Development Bank (ADB) 2021 *Climate Risk Country Profile: Malaysia*
[14] Mustafa M and Ariffin M 2011 Protection of Marine Biodiversity from Pollution: Legal Strategies in Malaysia. *Int. J. Biosci. Biochem. Bioinforma.* **1** 276-81
[15] ASEAN UP 2021 *8 free maps of ASEAN and Southeast Asia*
[16] Mohd-Zaki A H, Brett J, Ismail E and L'Azou M 2014 Epidemiology of Dengue Disease in Malaysia (2000-2012): A Systematic Literature Review *PLOS Negl. Trop. Dis.* **8** e3159
[17] Liang Y, Tan Q, Song Q. and Li J. 2021 An analysis of the plastic waste trade and management in Asia *Waste Manage* **119** 242-53
[18] Khazanah Research Institute 2019 *Plastic: An Undegradable Problem* p 16
[19] d'Ambrières W 2019 Plastics recycling worldwide: current overview and desirable changes. *Field Actions Sci. Rep.* **Special Issue 1 9 | 2019** 12-21
[20] Santodomingo N, Perry C, Waheed Z, Hussein M A B S, Rosedy A and Johnson K G 2021 Marine litter pollution on coral reefs of Darvel Bay (East Sabah, Malaysia) *Mar. Pollut. Bull.* **173** 112998
[21] Moh Y C and Manaf L A 2014 Overview of household solid waste recycling policy status and challenges in Malaysia *Resour Conserv Recy* **82** 50-61
[22] Dickella Gamaralalage P J, Pham N B, Nugroho S B, Devaporihatrakula C, Hngesbaugh M and Hayashi M 2020 *Plastic Waste in Southeast Asia - Urban Case Studies* United Nations Economic and Social Commission for Asia and the Pacific (UN-ESCAP)
[23] Fauziah S H, Rizman-Idid M, Cheah W, Loh K H, Sharma S, NoorMaiza M R, Bordt M, Praphotjanaporn T, Samah A A, Sabaruddin J S and George M 2021 Marine debris in Malaysia: A review on the pollution intensity and mitigating measures *Mar. Pollut. Bull.* **167** 112258

[24] Chen H L, Nath T K, Chong S, Foo V, Gibbons C and Lechner A M 2021 The plastic waste problem in Malaysia: management, recycling and disposal of local and global plastic waste *SN Appl. Sci.* **3** 437
[25] Bernama 2020 *SWCorp targets 40% recycling rate by 2025, current rate at 30%* TheStar
[26] Fauziah S H, Liyana I A and Periathamby A 2015 Plastic debris in the coastal environment: The invincible threat? Abundance of buried plastic debris on Malaysian beaches *Waste Manag. Res.* **33** 812-21
[27] Ibrahim N R and Noordin N N M 2020 Understanding the Issue of Plastic Waste Pollution in Malaysia: A Case for Human Security *Journal of Media and Information Warfare* **13** 105-40
[28] Ananthalakshmi A and Chow E 2018 Swamped with plastic waste: Malaysia struggles as global scrap piles up *Reuters*
[29] Kaur D 2018 *Plastic waste industry worth RM30b* The Malaysian Reserve
[30] Bernama 2019 *139 illegal plastic recycling plants ordered shut, says minister* Malay Mail
[31] Bernama 2019 *62 Malaysian companies hold permits to import plastic waste* New Straits Times
[32] Ministry of Energy, Technology, Science, Climate Change and Environment (MESTECC) 2018 *Malaysia's Roadmap Towards Zero Single-Use Plastics 2018-2030: Towards a sustainable future* p 16
[33] Shaharudin M S, Fernando Y, Ahmed E R and Shahudin F 2020 Environmental NOGs Involvement in Dismantling Illegal Plastic Recycling Factory Operations in Malaysia *Journal of Governance and Integrity* **4** 29-36
[34] Bujang J S, Zakaria M H and Short F T 2016 Seagrass in Malaysia: Issues and Challenges Ahead (*The Wetland Book* vol 1) ed C M Finlayson et al (Dordrecht: Springer Nature) p 9
[35] Asian Development Bank (ADB) 2014 *State of the Coral triangle: Malaysia* p 127
[36] International Coral Reef Initiative (ICRI) 2021 *Member Type Country: Malaysia*
[37] Nelms S E, Barnett J, Brownlow A, Davidson N J, Daeville R, Galloway T S, Lindeque P K, Santillo D and Godley B J 2019 Microplastics in marine mammals stranded around the British coast: ubiquitous but transitory? *Sci. Rep.* **9** 1075
[38] Lusher A 2015 Microplastics in the Marine Environment: Distribution, Interactions and Effects (*Marine Anthropogenic Litter)* M Bergmann et al (Bremerhaven: Springer Open) pp 245-307
[39] Anderson J C, Bradley J P and Palace V P 2016 Microplastics in aquatic environments: Implications for Canadian ecosystems *Environ. Pollut.* **218** 269-80
[40] Novillo O, Raga J A and Tomás J 2020 Evaluating the presence of microplastics in striped dolphins (*Stenella coeruleoalba*) stranded in the Western Mediterranean Sea. *Mar. Pollut. Bull.* **160** 111557
[41] Rahman S 2019 *A Risk Analysis of Microplastic Consumption in Filter Feeders* (Master's thesis, Nova Southeastern University, Florida, USA)
[42] Tang R 2018 *Anther Dead Dolphin Has Been Sighted In Penang In Less Than AWeek* SAYS
[43] Zakaria B K 2018 *Plastic thrown into oceans causing deaths of aquatic mammals* New Straits Times
[44] Cheznowski J A 2015 Distribution and Overlap of Five Coastal Indo-Pacific Cetacean Species *J. Aquac. Mar. Bio.* **2** 4 162-68
[45] Jaaman S A, Lah-Anyi Y U and Pierce G J 2008 Directed Fisheries for Dolphins and Dugongs in Sabah, East Malaysia: Past and Present *Borneo Science* **23**
[46] Bernama 2021 *Dolphin sightings in Kuala Perlis not unusual, says fisheries dept* Free Malaysia Today
[47] Inus K 2016 *Stranded dolphin died of stomach complications, says Sabah Wildlife Department* New Straits Times
[48] Ponnampalam L S 2012 Opportunistic observation on the distribution of cetaceans in the Malaysian South China, Sulu and Sulawesi seas and an updated checklist of marine mammals in Malaysia *Raffles Bull. Zool.* **60** 221-31
[49] Vanar M 2012 *Injured spinner dolphin rescued in Tuaran* TheStar

[50] Bono S, Kimura S S, Kuit S H, Ng J W, Tanaka K, Ichikawa K and Ponnampalam L S 2021 Description of the first acoustic recording of spinner dolphins (*Stenella longirostris*) from the northern straits of Malacca *Journal Acoust. Soc. Am.* **150** 2189
[51] Ponnampalam L S, Kuit S H & Ng J E 2018 Diversity and Distribution of Marine Mammals in Tun Mustapha Park, Kudat, Sabah, During the Tun Mustapha Park Scientific Expedition 2017 *Sabah Parks Nature Journal* **11** 25-46.
[52] Rodríguez-Vargas L H, Rajamani L, Dolar L and Porter L 2019 Population size, distribution and daylight behaviour of Irrawaddy dolphins (*Orcaella brevirostris*) in Penang Island, Malaysia *Raffles Bull. Zool.* **67** 671-83
[53] Rajamani L, Porter L, Dolar L, Rodríguez-Vargas L H and Yobe M 2018 Marine Mammals of Coastal Penang Island, Malaysia *Aquat. Mamm.* **44** 319-27
[54] Lajiun J 2015 *4.25kg of plastics in dolphin* Borneo Post Online
[55] Yang A Y, Ooi L S J, Phang S M, Hussin A H, Rosman F R and Fauzi R 2005 *Dugongs In Peril: The Conservation and Protection of Dugongs in Johor, Malaysia* (Kuala Lumpur, Report to the United Nations Development Programme)
[56] Ponnampalam L S, Izmal J H F, Adulyanukosol K, Ooi L S J and Reynolds J E III 2014 Aligning conservation and research priorities for proactive species and habitat management: the case of dugongs *Dugong dugon* in Johor, Malaysia *ORYX* **9** 743-49
[57] The Dugong & Seagrass Conservation Project 2021 *Malaysia: Country Overview*
[58] Rajamani L, Cabanban A S and Rahman R A 2006 Indigenous Use and Trade of Dugong (*Dugong dugon)* in Sabah, Malaysia *AMBIO* **35** 266-8
[59] Othman A F 2017 *Dugong found dead on Mersing beach, likely entangled in fishing net* New Strait Times
[60] Dangprasith P 2019 *Thailand's lost baby dugong dies from shock, eating plastic* AP News
[61] The IUCN Red List of Threatened Species 2021 Dugong: *Dugong dugon*
[62] Jolis G, Lau M M, Mustafa S R S, Sumampouw M, Rajan S G, Jumin R and Sharma D S K 2015 *Sea Turtle Conservation in Malaysia: Issues, challenges and recommendations* Seminar and Workshop on Sea Turtle Conservation in Malaysia, Kuala Terengganu, Terengganu, Malaysia
[63] The ASEAN Post team 2020 *Why are Malaysia's sea turtles disappearing?* THE ASEAN POST
[64] Bernama 2021 *Carcass of green turtle believed killed for its eggs found in Pulau Kapas* TheStar
[65] Bernama 2021 *Giant green turtle found dead without head and tail on Labuan beach* TheStar
[66] Bernama 2021 *Female green turtle found dead at Terengganu beach [NSTTV]* New Straits Times
[67] Riduan J 2021 *11 green turtles brutally killed in Semporna* New Straits Time
[68] Rahman R A, Wong E P, Joseph J, Salleh S M, Khoo, S N, Ismail M S, Rahim M A, Ramli R, Ang T L, Lee B H, Chan Z S and Long S L 2021 First Record of a Stranded Loggerhead Turtle (*Carette caretta*) in a Ghost Net off Penang, Malaysia *Marine Turtle Newsletter* **162** 22-4
[69] Liew H C 2017 *Conservation of Pacific Sea Turtles* ed Dutton P H Squires D and Ahmed M (Hawai'i: University of Hawai'i Press) pp 97-107
[70] Hamza A, Khir M A M, Rusli M U and Ibrahim Y I 2020 Microplastic Occurrence in Seaturtle Nesting Beach Sediments from Terengganu, Malaysia *J. Green Eng.* **10** 5712-29
[71] Alias N, Mansor M, Hussin M A, Husin T M, Azman N Z and Hassan N A 2020 Phenological Study of Mangrove Species on the West Coast Area of Peninsular Malaysia (*Status Mangroves in Malaysia)* ed H Omar et al (Kepong: Forest Research Institute Malaysia) chapter 5 pp 85-94
[72] Zakaria R M and Sharma S 2020 Mangrove Functions and Services: Roles of Carbon Pool in Climate Change Mitigation (*Status Mangroves in Malaysia)* ed H Omar et al (Kepong: Forest Research Institute Malaysia) chapter 2 pp 43-50
[73] Zakaria M and Rajpar M N 2015 Assessing the Fauna Diversity of Marudu Bat Mangrove Forest, Sabah, Malaysia, for Future Conservation *Diversity* **7** 137-48
[74] Polidoro B A et al The Loss of Species: Mangrove Extinction Risk and Geographic Areas of Global Concern *PLoS ONE* **5** e10095

[75] Omar H and Misman M A 2020 Extents and Distribution of Mangroves in Malaysia (*Status Mangroves in Malaysia)* ed H Omar et al (Kepong: Forest Research Institute Malaysia) chapter 1 pp 1-42
[76] Latiff A 2012 Conservation strategies for endangered mangrove swamp forest in Malaysia *Pak. J. Bot.* **44** 27-36
[77] See C M, Wong C H and Ng W P 2020 Mangrove Birds of North Central Selangor Coast, Selangor, Malaysia (*Status Mangroves in Malaysia)* ed H Omar et al (Kepong: Forest Research Institute Malaysia) chapter 13 pp 215-231
[78] The IUCN Red List of Threatened Species (2021, November 8). Grey-tailed Tattler: *Tringa brevipes*. https://www.iucnredlist.org/species/22693289/93394897
[79] The IUCN Red List of Threatened Species (2021, November 8). Spotted Greenshank: *Tringa guttifer*. https://www.iucnredlist.org/species/22693225/93391729
[80] The IUCN Red List of Threatened Species (2021, November 8). Oriental Darter: *Anhinga melanogaster*. https://www.iucnredlist.org/species/22696712/93582012
[81] Tangah J, Chung Y C A, Nilus R, Abdullah S Z S, Abu S and Lohuji P L 2020 Management System of Mangroves in Sabah: Issues and Challenges (*Status Mangroves in Malaysia)* ed H Omar et al (Kepong: Forest Research Institute Malaysia) chapter 9 pp 142-155
[82] Wilcox C, Sebille E V and Hardesty B D 2016 Threat of plastic pollution to seabirds is global, pervasive, and increasing *Proc. Natl. Acad. Sci. U. S. A.* **112** (38) 11899-04
[83] Ibrahim Y S, Azmi A A, Shukor S A, Anuar S T and Abdullah, S. A. 2016 Microplastics Ingestion by *Scapharca cornea* at Setiu Wetland, Terengganu, Malaysia *Middle-East J. Sci. Res.* **24** 2129-36
[84] Ibrahim Y S, Rathnam R, Anuar S T and Khalik W M A W M 2017 Isolation and Characterisation of Microplastic Abundance in *Lates calcarifer* from Setiu Wetlands, Malaysia *Malaysian J. Anal. Sci.* **21** 1054-64
[85] Hamid F S, Wang J and Zakaria R M 2020 Microplastics Abundance and Uptake by *Meretrix lyrate* (Hard Clam) in Mangrove Forest *J. Eng. Technol. Sci.* **52** 436-58
[86] Zaki M R M, Zaid S H M, Zainuddin A H and Aris A Z 2021 Microplastic pollution in tropical estuary gastropods: Abundance, distribution and potential sources of Klang River estuary, Malaysia *Mar. Pollut. Bull.* **162** 111866
[87] Hamzah S R, Altrawneh R S, Anuar S T, Khalik W M A W M, Kolandhasamy P and Ibrahim Y S 2021 Ingestion of microplastics by the estuarine polychaete, *Namalycastis* sp. in the Setiu Wetlands, Malaysia *Mar. Pollut. Bull.* **170** 112617
[88] Barasarathi J, Periathamby A, Fauziah S H and Emenike C 2014 *Microplastic Abundance in Selected Mangrove Forest in Malaysia* Proceedings of The ASEAN Conference on Science and Technology 2014 (Bogor, Indonesia)
[89] Lim E V, Izam N I I, Nilamani N, Razalli N M, Zanuri N M, Yasin Z, Tan S H A, Zhang S and Li H 2020 Abundance and distribution of macro- and microplastics at three different habitats along the Penang Coastline in the Northern Straits of Malacca *Ramkhamhaeng International Journal of Science and Technology* **3**
[90] Ibrahim Y S, Hamzah S R, Khalik W M A W M, Yusof K M K K and Anuar S T 2021 Spatiotemporal microplastic occurrence study of Setiu Wetland, South China Sea. *Sci. Total Environ.* **788** 147809
[91] Zaki M R M, Peh X Y, Zainuddin A H, Razak M R and Aris A Z 2021 Occurrence, abundance, and distribution of microplastics pollution: an evidence in surface tropical water of Klang River estuary, Malaysia *Environ. Geochem. Health* **43** 1 3733-48
[92] Rabbani G, Yan B C, Li Y L N, Ooi L S J, Lee J N, Huang D and Wainwright J 2021 Spatial and Structural Factors Shape Seagrass-Associated Bacterial Communities in Singapore and Peninsular Malaysia. *Front. Mar. Sci.* **8** 659180
[93] Rajamani L and Marsh H 2015 Mapping seagrass cost-effectively in the Coral Triangle: Sabah, Malaysia as a case study *Pac. Conserv. Biol.* **21** 113–21

[94] Ramesh R, Banerjee K, Pannerselvam A, Raghuraman R, Purvaja R and Lakshmi A 2019 Importance of Seagrass Management for Effective Mitigation of Climate Change *(Coastal Management)* ed R R Krishnamurthy et al (Massachusetts: Academic Press) chapter 14 pp 283-299
[95] Huang Y, Xiao X, Xu C, Perianen Y D, Hu J and Holmer M 2019 Seagrass beds acting as a trap of microplastics – Emerging hotspot in the coastal region? *Environ. Pollut.* **257** 113450
[96] Sanchez-Vidal A, Canals M, de Haan W P, Romero J and Veny M 2021 Seagrasses provide a novel ecosystem service by trapping marine plastics *Sci. Rep.* **11**
[97] Unsworth R K F, Higgs A, Walter B, Cullen-Unsworth L C, Inman I and Jones B L 2021 Canopy Accumulation: Are Seagrass Meadows a Sink of Microplastics? *Oceans* **2** 162-76
[98] Hashim U A B, Abdullah N and Takeshi M 2019 *A Review on Plastic Policies in Malaysia and Japan* The Japan Section of the Regional Science Association International
[99] Selan S 2021 *Plastic bag charge won't cut it, says activist in anti-pollution push* Malaysia Now
[100] Trisha N 2021 *'Plastic bags dearer but not better'* TheStar
[101] Bernama 2021 *'Say no to plastic straws' campaign seems to have lost its pull'* The Sun Daily

Analysis of the Characteristics of Heavy Metal Elements in Soil Around the Smelting Area of Pasuruan Industrial Estate, Indonesia

A G A Ammar[1], A D Syafei[1*], M Santoso[2], B V Tangahu[1], A F Assomadi[1] and J Hermana[1]

[1] Department of Environmental Engineering, Institut Teknologi Sepuluh Nopember, Surabaya
[2] Center of Nuclear Technology for Materials and Radiometry, BATAN, Bandung
*Corresponding author: dipareza@enviro.its.ac.id

Abstract. Smelting plant is an activity that produces emissions in the form of heavy metal particulates that can cause pollution and are harmful to human health in excessive concentrations. This research was conducted to determine the heavy metals' characteristics in the soil around the metal smelting area of Beji industrial area, Pasuruan, Indonesia. A total of 33 soil samples were collected at 200 m, 500 m, 2km and 5km from the metal smelting area. The concentrations of heavy metals Zn (669.47 - 63.25 mg/kg), Cd (94.88 - 40.30 mg/kg) and V (198.98 - 108.19 mg/kg) in the soil were determined by using EDXRF Spectrometer Minipal 4. The concentration of zinc reduces with distance from an industrial region. The amounts of cadmium and vanadium did not change considerably with distance. Then the potential pollution risks of heavy metal in the soil were evaluated by method of assessment of heavy metal contaminations in soil. These C_f values indicated that the assessment of heavy metal contamination in soils from smelters was category: Indicates no contamination, suspected contamination, slight contamination and extreme contamination.

1. Introduction

Natural and anthropogenic sources contribute to the discharge of heavy metals into the environment, elevated metal concentrations in the environment are a result of human activity [1]. Soil has been identified as the primary sink for anthropogenic heavy metal deposition through a variety of mechanisms [2-6]. Heavy metals and metalloids may accumulate in soils as a consequence of emissions from rapidly growing industrial regions, coal combustion residues, and atmospheric deposition [7-8]. Unlike organic pollutants which are oxidized by microbial activity into carbon (IV) oxide, most metals are not degraded by microbial or chemical action [9], and their cumulative concentration in soils remains for an extended period after their introduction [10]. However, modifications to their chemical forms (speciation) and bioavailability are conceivable. Toxic metals in soil may significantly impair the biodegradation of organic pollutants [11]. Metal polluted soil presents a danger to people and animals through consumption of plants that contain bioaccumulated contaminated soil hazardous metals [12]. The measurement of heavy metal levels in soil is a crucial indication of anthropogenic input in the environment due to the dangers presented by heavy metals in soil [13-14].

This paper studies the heavy metal contamination in the soil of Pasuruan, Indonesia's Beji metal smelting region. The assessment of heavy metal contamination in soil was used to ascertain the amount of Zn, Cd, and V contamination. This research aimed to determine the concentration of heavy metals in the soil around a metal smelting region and provide a scientific foundation for heavy metal pollution management.

2. Material and Methods

2.1 Sampling Site

Soil samples were taken in the Beji industrial area, Pasuruan, Indonesia. This area's climate is classified as tropical, with annual averages of humidity, temperature, and rainfall of 77, 12%, 27.7°C, and 734 mm, respectively. The sampling sites were shown in Figure 1. A soil sampling standard developed by the United States Environmental Protection Agency (US EPA) was adopted [15]. All experimental procedures relate to the ISO/IEC 17025:2005 quality assurance standard, which is applied at the BATAN laboratory.

Figure 1. Location of the study area in Pasuruan, Indonesia.

2.2 Sample Collection

Soil samples were collected at 200 m, 500 m, 2km and 5km from the metal smelting area at an approximate depth of 0-15 cm, 20-30 cm and 35-40 cm. A total of 33 samples were collected in the west, northeast, east northeast, east southeast and south southeast of the Industrial. At every sampling point, coordinates were taken by a handheld GPS (Garmin GPSMAP 64s). All soil samples were then placed in polyethene bags, were given a laboratory number and transported to the laboratory.

2.3 Sample preparation

Soil sample preparation was carried out in the laboratory. Soil samples were placed on a plastic tray, cleaned of impurities such as leaves, stones, twigs, etc., and then dried in an oven. After the drying process, the lumps and aggregates in the soil sample were crushed and ground using a glass pestle and mortar, and then the samples were ground and sieved to obtain a particle size of 150 mesh. The next step is to prepare the sample container for the analysis process. The container used is made of plastic which is covered with mylar film underneath. Then the soil sample was put into a container and weighed using a Mettler Toledo Micro Balance Type ME2002E and then compacted.

2.4 Sample measurement
Around 1 gram of non-pressed standard reference material (SRM) or soil samples were put in plastic sample containers then fed into the EDXRF Spectrometer Minipal 4. The duration of the measurement was 300 seconds.

2.5 Assessment of heavy metal contaminations in soil
About the potential ecological hazard of heavy metal pollution in surface soils was evaluated using the ecological risk index introduced by Hakanson, 1980 using Equation 1:

$$C_{fi} = C_i / C_{ref}$$
(Equation 1)

where, C_{fi} = contamination factor/index; C_i = metal concentrations at each sampling point, C_{ref} = evaluation criterion of the metal [16].

Table 1. Index of Contamination and Contamination Level

C_f	Contamination Level
1	Indicates no contamination
1- 2	Suspected contamination
2-3.5	Slight contamination
3.5 - 8	Moderate contamination
8-27	Severe contamination
>27	Extreme contamination

Gonzales and Miqueo classified contamination into six different categories [17]. The contamination index and the level of contamination are shown in Table 1.

3. Result and discussion
Heavy metal concentrations of Zn, Cd, and V based on depth and distance, as well as the contamination factors of heavy metal in soil investigated in this study are shown in Table 2-4. The elemental zinc concentration in soil samples at a depth of 0 - 15 cm (64.63 - 275.54 mg/kg), 20 - 30 cm (63.77 - 669.47 mg/kg) and at a depth of 35 - 40 cm (63.84 - 189.03 mg/kg). Cadmium concentrations in soil samples were determined at depths of 0 - 15 cm (40.30 - 94.88 mg/kg), 20 - 30 cm (41.57 - 87.93 mg/kg), and 35 - 40 cm (44.25 - 90.95 mg/kg). Vanadium concentrations in soil samples at 0 - 15 cm (117.40 - 189.40 mg/kg), 20 - 30 cm (125.37 - 178.28 mg/kg), and 35 - 40 cm (108.19 - 180.63 mg/kg). It can be found that the concentration of zinc decreases with increasing distance from the industrial area; it can be seen at a distance of 200 meters has a concentration value nine times greater than a distance of 5 km from the industry. On the other hand, the concentrations obtained for cadmium and vanadium did not significantly differ with increasing distance. Calculation of heavy metal Cf in soil samples is shown in Figure 2.

Table 2. Summary of concentration, depth and distance ratio of zinc in soil samples

Distance	Direction	Concentration (mg/kg)		
		0 - 15 cm	20 - 30 cm	35 - 40 cm
200 meter	West	275.54	669.47	176.82
	East Northeast	118.21	81.15	79.81
	East Southeast	77.47	73.99	76.34
500 meter	West	86.37	74.62	68.67
	East Northeast	238.43	203.20	189.03
	East Southeast	77.48	67.16	63.84
2 km	West	84.34	73.03	74.54
	East Northeast	96.36	71.48	68.72
	South Southeast	64.63	63.77	63.25
5 km	Northeast	71.79	71.20	78.94
	East Northeast	73.97	70.60	71.25

Table 3. Summary of concentration, depth and distance ratio of cadmium in soil samples

Distance	Direction	Concentration (mg/kg)		
		0 - 15 cm	20 - 30 cm	35 - 40 cm
200 meter	West	74.04	86.87	73.48
	East Northeast	94.88	87.93	90.95
	East Southeast	64.90	64.77	77.19
500 meter	West	52.82	58.61	66.36
	East Northeast	52.19	51.54	44.25
	East Southeast	40.30	50.34	44.71
2 km	West	49.95	49.90	56.04
	East Northeast	56.66	52.42	51.91
	South Southeast	49.58	41.57	50.33
5 km	Northeast	44.68	60.16	51.10
	East Northeast	49.93	47.38	47.68

Table 4. Summary of concentration, depth and distance ratio of vanadium in soil samples

Distance	Direction	Concentration (mg/kg)		
		0 - 15 cm	20 - 30 cm	35 - 40 cm
200 meter	West	153.76	178.28	180.63
	East Northeast	198.98	157.55	149.26
	East Southeast	154.46	150.42	144.65
500 meter	West	177.98	178.76	167.14
	East Northeast	138.22	137.38	137.43
	East Southeast	135.55	137.84	133.82
2 km	West	145.75	137.97	132.91
	East Northeast	138.16	138.40	141.48
	South Southeast	117.40	125.37	108.19
5 km	Northeast	132.47	132.59	142.15
	East Northeast	189.40	167.64	164.70

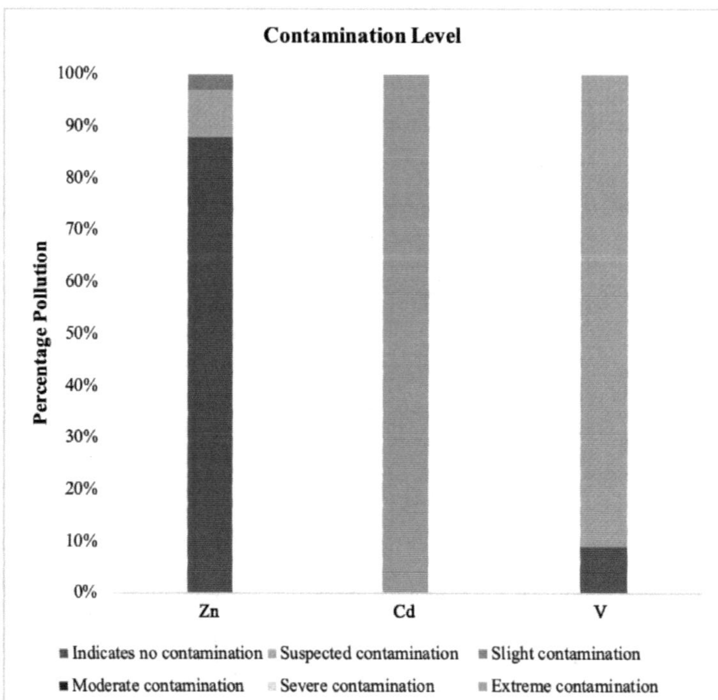

Figure 2. Percentage C_f of Zn, Cd, and V in soil samples

These C_f values indicated that the assessment of heavy metal contamination in soils from smelters was category: Indicates no contamination, suspected contamination, slight contamination and extreme contamination. For Zn, 87.88% in category indicates no contamination, 9.09% in category suspected contamination, 3.03% in category slight contamination. For Cd, 100% sample in extreme contamination. For V, 9.1% in category indicates no contamination, 90.9% in category suspected contamination.

4. Conclusion

The concentration of zinc decreases with increasing distance from the industrial area; it can be seen at a distance of 200 meters has a concentration value nine times greater than a distance of 5 km from the industry. On the other hand, the concentrations obtained for cadmium and vanadium did not significantly differ with increasing distance. As indicated by this C_f value, the assessment of heavy metal contamination in soil from smelters is classified into the following categories: Indicates no contamination, suspected contamination, slight contamination, and extreme contamination.

References

[1] Jung M.C. 2001 Heavy metal contamination of soils and waters in and around the Incheon Au-Ag mine, Korea. *Applied Geochem*; **16**:1369–1375.

[2] Harrison, R.M., Laxen, D.P.H., Wilson, S.J., 1981. Chemical associations of lead, cadmium, copper, and zinc in street dusts and roadside soils. Environmental Science & Technology **15**, 1378–1383.

[3] Li, X.D., Poon, C.S., Liu, P.S., 2001. Heavy metal contamination of urban soils and street dusts in Hong Kong. Applied Geochemistry **16**, 1361–1368

[4] Al–Khashman, O.A., Shawabkeh, R.A., 2006. Metals distribution in soils around the cement factory in southern Jordan. Environmental Pollution 140, 387–394.

[5] Rashed, M.N., 2010. Monitoring of contaminated toxic and heavy metals, from mine tailings through age accumulation, in soil and some wild plants at Southeast Egypt. Journal of Hazardous Materials **178**, 739–746.

[6] Guo, G., Wu, F., Xie, F., Zhang, R., 2012. Spatial distribution and pollution assessment of heavy metals in urban soils from southwest China. Journal of Environmental Sciences **24**, 410–418.

[7] S. Khan, Q. Cao, Y. M. Zheng, Y. Z. Huang, and Y. G. Zhu. 2008. “Health risks of heavy metals in contaminated soils and food crops irrigated with wastewater in Beijing, China,” Environmental Pollution, vol. 152, no. 3, pp. 686–692,

[8] M. K. Zhang, Z. Y. Liu, and H.Wang. 2010. “Use of single extraction methods to predict bioavailability of heavy metals in polluted soils to rice,” Communications in Soil Science and Plant Analysis, vol. 41, no. 7, pp. 820–831.

[9] T. A. Kirpichtchikova, A.Manceau, L. Spadini, F. Panfili,M. A.Marcus, and T. Jacquet. 2006. “Speciation and solubility of heavy metals in contaminated soil using X-ray microfluorescence, EXAFS spectroscopy, chemical extraction, and thermodynamic modeling,” Geochimica et Cosmochimica Acta, vol. 70, no. 9, pp. 2163–2190.

[10] D. C. Adriano. 2003.Trace Elements in Terrestrial Environments: Biogeochemistry, Bioavailability and Risks of Metals, Springer, New York, NY, USA, 2nd edition.

[11] P. Maslin and R. M. Maier. 2000. “Rhamnolipid-enhanced mineralization of phenanthrene in organic-metal co-contaminated soils,” Bioremediation Journal, vol. 4, no. 4, pp. 295–308

[12] Turner, A.H. 2009. Urban Agriculture and Soil Contamination: An Introduction to Urban Gardening, Environmental Finance Centre, EPA Region 5, University of Louisville, pp. 1–17.

[13] Manta, D.S., Angelone, M., Bellanca, A., Neri, R., Sprovieri, M. 2002. Heavy metals in urban soils: A case study from the city of Palermo (Sicily), Italy. Science of the Total Environment **300**, 229–243

[14] Addo, M.A., Darko, E. O., Gordon, C., Nyarko, B.J.B., Gbadago, J. K., Nyarko, E., Affum, H. A., Botwe, B.O. 2012. Evaluation of heavy metals contamination of soil and vegetation in the vicinity of a cement factory in the Volta region, Ghana. International Journal of Science and Technology **2**, 40–50.

[15] US EPA SESDPROC-300-R3. 2014. Operating Procedur Soil Sampling Effective Date: Aug 21

[16] Hakanson, L. 1980. An ecological risk index for aquatic pollution–control – a sedimentological approach. Water Research **14**, 975–1001.
[17] Gonzalez-Miqueo, L., Elustondo, D., Lasheras, E., Santamaria, J.M. 2010. Use of native mosses as biomonitors of heavy metals and nitrogen deposition in the surrounding of two steel works. Chemosphere **78**, 965-971.

Acknowledgments
The author would like to thank the Department of Environmental Engineering, Faculty of Civil, Environmental and Geo-Engineering, Institut Teknologi Sepuluh Nopember (ITS) for the support and encouragement of this research. We want to thank the Center of Nuclear Technology for Materials and Radiometry, BATAN, Bandung for helped the sample preparation and analysis of data.

A Review of Plastic-derived Diesel Fuel as a Renewable Fuel for Internal Combustion Engines: Applications, Challenges, and Global Potential

Wan Nurdiyana Wan Mansor[1,2,*], Nurul Ashraf Razali[1,2], Samsuri Abdullah[1], Mohammad Nor Khasbi Jarkoni[1], Anis Busrya Eddy Sharin[1], Nurul Huda Abd Kadir[2,3], Aima Ramli[2,3], How-Ran Chao[4], Sheng-Lun Lin[5], Juliana Jalaludin[6]

[1] Faculty of Ocean Engineering Technology & Informatics, Universiti Malaysia Terengganu, Malaysia
[2] Syngas Research Interest Group, Universiti Malaysia Terengganu, Malaysia
[3] Faculty of Science and Marine Environment, Universiti Malaysia Terengganu, Malaysia
[4] Department of Environmental Science and Engineering, National Pingtung University of Science and Technology, Taiwan
[5] School of Mechanical Engineering, Beijing Institute of Technology, China
[6] Department of Environmental and Occupational Health, Faculty of Medicine and Health Science, University Putra Malaysia, Malaysia
*Corresponding author email address: nurdiyana@umt.edu.my

Abstract. Plastics' broad application in a wide range of industries has resulted in an increase in global plastic production rates over time. The continued need for plastic, which can take up a lot of space in landfills due to the accumulation of plastic trash, has contributed to environmental problems. To address the problem of plastic waste, many options, such as recycling and energy recovery technologies, have been created. One possible technique to address landfill issues is to convert plastic waste into diesel fuel. In this review, the potential use of plastic-derived diesel fuel in internal combustion engines as a renewable and sustainable energy source is discussed. It's been found that due to similar properties in plastic-derived diesel fuel, it inherits similar performance in internal combustion engines.

1. Introduction

Plastic has risen to become one of the most widely utilized materials. It is less expensive than alternative materials such as paper and fabric. Because of its low cost, plastic is a popular material for a wide range of applications. It has been utilized in a wide range of activities including healthcare, construction, packaging, and engineering [1-3]. Polyethylene Terephthalate (PET or PETE), High-Density Polyethylene (HDPE), Polyvinyl Chloride (PVC), Low-Density Polyethylene (LDPE), Polypropylene (PP), Polystyrene or Styrofoam (PS), and miscellaneous plastics are the seven types of plastic that are most common in the modern world today [4,5]. The rapid expansion of the world's population has increased the need for commodity plastics. More than 8.3 billion tonnes of plastics were manufactured worldwide between 1950 and 2021 but only 9% of that plastic was recycled, 12% was incinerated, while the remaining 79 % was discarded in the environment [6]. In Malaysia, 19% of the total garbage was made up of plastic in 2007 [7]. Decomposing plastic in landfills might take up to 500 years, and it could release contaminants into surrounding soil and water if not handled properly, which could have adverse effects on the environment and human health. Previous studies showed that endocrine disruptors, malignancies, developmental disorders, hypertension and reproductive abnormalities have all been linked to plastics' chemical compounds and additives such as phthalates, bisphenol A or polybrominated diphenyl ethers [8-10]. Plastic also breaks down into microscopic plastics that pollute the environment and marine life. Because of their tiny size, microscopic plastics are easily ingested by a wide variety of marine species, ranging from high- to low-trophic level organisms. It impairs their growth, metabolism, reproduction, and cellular response [11-14]. It is essential to properly dispose of plastic waste and transform it into renewable energy sources. Since plastic is mostly made of carbon and hydrogen, it has the same amount of energy as diesel or other traditional fuels. Pyrolysis is a process for converting plastic waste into energy in the form of solid, liquid, and gaseous fuels [15,16]. In pyrolysis, plastic waste is heated to a high temperature in the absence of oxygen to break down it. This study aims to discuss the properties of plastic-derived diesel fuel to determine their suitability as a fuel. Additionally, the application, problems and global potential associated with the utilization of waste plastic diesel fuel in a diesel engine will be examined.

2. Physical and chemical properties of plastic-derived diesel fuel relevant to engines

Ashes and volatile matter are the key determinants of liquid oil output in the pyrolysis process. High volatile matter encouraged the production of liquid oil, whereas high ash content reduced it, increasing gaseous yield and char formation [17]. The investigation is performed using a proximate analysis. The proximate analysis refers to the process of determining the ash content, moisture content, volatile matter, and fixed carbon content of a sample. Table 1 shows the proximate analysis of several polymers. Table 2 summarises several of the properties of liquid fuel produced by pyrolysis of mixture/single polymer as reported in the literature.

Table 1. Proximate analysis of plastics [18-24].

Types of plastics	Plastic type marks	Moisture (wt%)	Fixed carbon (wt%)	Volatile (wt%)	Ash (wt%)
PET	1 PET	0.46	7.77	91.75	0.02
HDPE	2 HDPE	0.00	0.01	99.81	0.18
PVC	3 V	0.80	6.30	93.70	0.00
LDPE	4 LDPE	0.30	0.00	99.70	0.00
PP	5 PP	0.15	1.22	95.08	3.55
PS	6 PS	0.25	0.12	99.63	0.00
Polyethylene (PE) Acrylonitrile butadiene styrene (ABS)	7 OTHER	0.10	0.04	98.87	0.99
Polyamide (PA) or Nylons Polybutylene terephthalate (PBT)		0.00	0.69	99.78	0.00

Table 2. Fuel properties

Type of plastic	Mixture of HDPE, LDPE, PP, PS, PET	PE, PS and 70% organic matter	Mainly ABS, PS and clay	Virgin PP	Waste PP	Diesel
Density@15-15.6°C (kg/l)	0.743	0.825	0.9813	0.768	0.769	0.8398
Kinematic viscosity@40°C (cSt)	3.5	3.76	1.918	-	-	2.62
Flash point (°C)	47	41	13	-	-	59.5
Fire point (°C)	53	-	-	-		-
Pour point (°C)	- 4	-	-	2.7	- 4.7	-
Cloud point (°C)	-36	-	-	-	-	-
Calorific Value (MJ/kg)	41.1	40.58	38.3	45.63	46.04	42.9
Aromatic content (%)	29	-	65.5	-	-	29.5
Reference	[25]	[26]	[27]	[28]	[28]	[27]

3. Engine performance and emissions using plastic-derived diesel fuel

Engine performance is a measure of the engine's ability to complete its designated function, which is the conversion of the chemical energy contained in the fuel to meaningful mechanical work. The following factors are used to determine an engine's performance: combustion pressure, brake thermal efficiency (BTE), brake specific fuel consumption (BSFC). The pressure is a measure of the power developed by the engine, BTE measures the effectiveness of the conversion from the chemical energy contained in the fuel (input) into mechanical work and BSFC is the measurement of fuel efficiency by an engine that combusts the air-fuel mixture and generates the rotational motion of a crankshaft through combustion. In perfect combustion of a diesel engine, only energy, carbon dioxide (CO_2), and water vapor (H_2O) would be produced. However, as a result of several non-ideal combustion processes, diesel engines create nitrogen oxides (NOx), unburned hydrocarbons (UHC), carbon monoxide (CO), and hazardous particulates. Table 3 summarizes the performance and exhaust emission characteristics of CI engines fuelled with diesel-derived plastic fuel in comparison with diesel fuel. As shown, the peak pressure of diesel engines with plastic oil is increased or similar to diesel fuel. Plastic oil caused a longer ignition delay in the premixed combustion phase. This is due to the high viscosity of plastic oil. The in-cylinder pressure is increased because of a longer ignition delay in waste plastic oil [29,30]. Additionally, the high viscosity, poor atomization, heavier hydrocarbons, and bigger aromatic bonds resulted in the chain requiring more energy to break down [29-35]. As a result, the BTE decreased and BSFC increased. Exhaust emissions were reported to increase for all types of plastic fuel. The increase in NOx might be attributed to the increased exhaust gas temperature of plastic oil, which results in more oxygen atoms combined. This increase is also a result of the high cylinder pressure and ignition delay, which results in a greater amount of heat being released when compared to other test fuels [36]. Another possibility is that the presence of more aromatic compounds in plastic oil raises the adiabatic flame temperature, resulting in a rapid rate of heat release and a high level of NOX in the exhaust [37]. The higher CO emission in plastic oil is a result of its increased viscosity, which inhibits fuel vaporization and atomization, hence promoting incomplete combustion of the air-fuel mixture. The higher CO concentration in plastic oil than in diesel is also most likely due to the low oxygen component of the fuel [38]. The increased UHC emission from plastic oil is owing to its high density and viscosity, which prevent it from burning completely throughout the combustion chamber, exiting as an incomplete combustion product [39].

Table 3. Summary of performance and exhaust emission characteristic of CI engines fuelled with diesel-derived plastic fuel in comparison with diesel fuel.

Type of plastic fuel	Calorific value Diesel/Plastic Oil (MJ/kg)	Engine type	Peak Pressure	BTE	BSFC	NOx	UHC	CO	Ref.
HDPE	43.5/44.2	Single cylinder, 4-stroke, water-cooled, naturally aspirated DI diesel engine	Increased	Decreased	-	Increased	Increased	Increased	[29]
-	43.9/44.6	Single-cylinder, water-cooled DI diesel engine	-	Decreased	Increased	Increased	Increased	Increased	[30]
HDPE+PET+LDPR+PVC+PP+PS	43.3/44.4	Single-cylinder with 4-stroke DI diesel engine	Increased	Decreased	Increased	Increased	Increased	Increased	[31]
HDPE	43.5/44.2	Single cylinder, 4-stroke, water-cooled, DI diesel engine,	-	Decreased	Increased	Increased	Increased	Increased	[32]
waste plastic	43.5/44.2	Single cylinder, 4-stroke, water-cooled, DI diesel engine	Increased	Decreased	-	-	-	-	[33]
LDPE and polybags	45.5/40.4	Single-cylinder, liquid-cooled, naturally aspirated diesel engine	Similar	Decreased	Increased	Increased	Increased	Increased	[34]
-	46.5/45.2	Single-cylinder, water-cooled, 4 stroke diesel engine	-	Decreased	Increased	Similar	increased	Increased	[35]

4. Global potential and challenges

Pyrolysis is gaining attention as a method of converting plastic to liquid fuel due to its storage and transportation advantages, as well as its applicability to applications such as combustion engines, boilers, and turbines [40, 41]. The calorific value of plastic oil is comparable to that of conventional diesel. As a result, it has the potential to be used as a source of alternative energy and a mode of transportation. Additionally, it can be blended with diesel, biodiesel fuel, and additives to improve performance and efficiency [42]. The global annual production of plastic has been estimated at 300 million tonnes, and this figure is rapidly increasing [43]. Pyrolysis of plastic waste provided a high output of liquid fuel of more than 80% by weight, according to a prior study [44,45]. As a result, the plastic oil sector currently has the potential to produce 240 million tonnes of plastic oil annually. When compared to generating the same amount of polymers from virgin fossil fuel, recycling every additional tonne of plastic reduces emissions by 1.1–3.0 tonnes of CO_2e [46]. Despite the decreased engine performance of plastic pyrolysis oil, the amount of plastic waste generated is large, and it must be handled to avoid environmental hazards. Not only is converting waste to energy a financially advantageous investment, but it is also an environmentally friendly option.

A major drawback of the pyrolysis process is that it is energy expensive, and the resulting oil requires further energy to refine. Another key issue is the lack of readily available plastic feedstocks for pyrolysis, due to the inefficiency of recycling at the moment, such as mechanical separation, pelletization, and regeneration, including high labor costs and water contamination, particularly in developing countries [47]. The pyrolysis emission treatment technique must be further enhanced to achieve maximum environmental benefits due to harmful gases such as sulfur, created by the pyrolysis of some plastic waste, such as HDPE and mixed plastics [48,49]. To ensure the lowest environmental impact, the pyrolysis oil derived from various plastic types must be cleaned extensively before it is utilized in any application [50]. Aromatic hydrocarbons, some of which are well-known carcinogens, have the potential to do significant harm to human health and the environment [51]. This necessitates serious consideration.

5. Conclusions

Pyrolysis is an innovative method for converting plastic waste to liquid oil. Plastic oil derived from several forms of plastic waste showed comparable heating values to regular diesel, making it appropriate for use in diesel engines. In comparison to traditional diesel, it has lower thermal efficiency and a greater BSFC. Additionally, as compared to diesel fuel, plastic fuel generates greater levels of NOx, CO, and UHC. Additionally, the pyrolysis of plastic waste into diesel fuel has some technical, operational, and post-process challenges that must be solved to maximise economic and environmental advantages.

References

[1] Al-salem, S.M., Antelava, A., Constantinou, A., Manos., A, Dutta, A 2017 A review on thermal and catalytic pyrolysis of plastic solid waste *Journal of Environmental Management* **197**, 177-198.

[2] J. Lange, Y. Wyser 2003 Recent innovations in barrier technologies for plastic packaging—a review *Packag. Technol Sci* **16** (4), pp. 149-158

[3] Jasmin Wiefek, Julia Steinhorst, Katharina Beyerl 2021 Personal and structural factors that influence individual plastic packaging consumption—Results from focus group discussions with German consumers *Cleaner and Responsible Consumption* Volume **3** 100022,

[4] Bassam A. Tayeh, Ibrahim Almeshal, Hassan M. Magbool, Hisham Alabduljabbar, Rayed Alyousef 2021 Performance of sustainable concrete containing different types of recycled plastic *Journal of Cleaner Production* Volume **328** 129517

[5] Sunantha Ganesan, Thanaporn Ruendee, Susana Y. Kimura, Chamorn Chawengkijwanich, Dao Janjaroen 2022 Effect of biofilm formation on different types of plastic shopping bags: Structural and physicochemical properties *Environmental Research* Volume **206** 112542

[6] Zaman, A., Newman, P 2021 Plastics: are they part of the zero-waste agenda or the toxic-waste agenda? *Sustain Earth* **4**, 4 https://doi.org/10.1186/s42055-021-00043-8

[7] Wahab DA, Abidin A, Azhari CH 2007 Recycling trends in the plastics manufacturing and recycling companies in Malaysia. *J Appl Sci* **7**(7):1030–1035

[8] Trasande, L., R. T. Zoeller, U. Hass, A. Kortenkamp, P. Grandjean, J. P. Myers, J. DiGangi, M. Bellanger, R. Hauser, J. Legler, N. E. Skakkebaek, and J. J. Heindel 2015 Estimating burden and disease costs of exposure to endocrine-disrupting chemicals in the European Union *Journal of Clinical Endocrinology & Metabolism* **100**: 1245–55.

[9] Darbre, PD. Chemical components of plastics as endocrine disruptors: Overview and commentary 2020 *Birth Defects Research* **112**: 1300– 1307. https://doi.org/10.1002/bdr2.1778

[10] Okugbe E. Ohore, Songhe Zhang, Endocrine disrupting effects of bisphenol A exposure and recent advances on its removal by water treatment systems. A review, Scientific African, Volume 5, 2019, e00135, ISSN 2468-2276, https://doi.org/10.1016/j.sciaf.2019.e00135.

[11] Mao Y., Ai H., Chen Y., Zhang Z., Zeng P., Kang L., Li W., Gu W., He Q., Li H. 2018 Phytoplankton response to polystyrene microplastics: Perspective from an entire growth period *Chemosphere* **208**:59–68. doi: 10.1016/j.chemosphere.2018.05.170

[12] Sjollema S.B., Redondo-Hasselerharm P., Leslie H.A., Kraak M.H.S., Vethaak A.D. 2016 Do plastic particles affect microalgal photosynthesis and growth? *Aquat.Toxicol* **170**:259–261. doi: 10.1016/j.aquatox.2015.12.002

[13] Lu, I-Cheng, How-Ran Chao, Wan-Nurdiyana-Wan Mansor, Chun-Wei Peng, Yi-Chyun Hsu, Tai-Yi Yu, Wei-Hsiang Chang, and Lung-Ming F. 2021 Levels of Phthalates, Bisphenol-A, Nonylphenol, and Microplastics in Fish in the Estuaries of Northern Taiwan and the Impact on Human Health *Toxics* **9**, no. 10: 246. https://doi.org/10.3390/toxics9100246

[14] Martinez-Gomez C., Leon V.M., Calles S., Gomariz-Olcina M., Vethaak A.D 2017 The adverse effects of virgin microplastics on the fertilization and larval development of sea urchins. *Mar. Environ. Res* **130**:69–76. doi: 10.1016/j.marenvres.2017.06.016

[15] Senthil Kumar, P., Bharathikumar, M., Prabhakaran, C. et al. 2017 Conversion of waste plastics into low-emissive hydrocarbon fuels through catalytic depolymerization in a new laboratory scale batch reactor *Int J Energy Environ Eng* **8**, 167–173 https://doi.org/10.1007/s40095-015-0167-z

[16] Miao, Y.; von Jouanne, A.; Yokochi, A.2021 Current Technologies in Depolymerization Process and the Road Ahead *Polymers* **13**, 449. https://doi.org/10.3390/polym13030449

[17] F. Abnisa, W.M.A. Wan Daud 2014 A review on co-pyrolysis of biomass: an optional technique to obtain a high-grade pyrolysis oil *Energy Convers Manage* **87** pp. 71-85

[18] F. Zannikos, S. Kalligeros, G. Anastopoulos, E. Lois 2013 Converting biomass and waste plastic to solid fuel briquettes *J Renew Energy* Volume **2013** http://dx.doi.org/10.1155/2013/360368

[19] I. Ahmad, M. Ismail Khan, M. Ishaq, H. Khan, K. Gul, W. Ahmad 2013 Catalytic efficiency of some novel nanostructured heterogeneous solid catalysts in pyrolysis of HDPE *Polym Degrad Stab* **98** pp. 2512-2519

[20] S.-J. Hong, S.C. Oh, H.-P. Lee, H.T. Kim, K.-O. Yoo 1999 A study on the pyrolysis characteristics of poly(vinyl chloride*)* *J Korean Inst Chem Eng* **37** pp. 515-521

[21] S.S. Park, D.K. Seo, S.H. Lee, T.-U. Yu, J. Hwang 2012 Study on pyrolysis characteristics of refuse plastic fuel using lab-scale tube furnace and thermogravimetric analysis reactor *J Anal Appl Pyrol* **97** pp. 29-38

[22] S.-H. Jung, M.-H. Cho, B.-S. Kang, J.-S. Kim 2010 Pyrolysis of a fraction of waste polypropylene and polyethylene for the recovery of BTX aromatics using a fluidized bed reactor *Fuel Process Technol* **91** pp. 277-284

[23] F. Abnisa, W.M.A.W. Daud, J.N. Sahu 2014 Pyrolysis of mixtures of palm shell and polystyrene: an optional method to produce a high-grade of pyrolysis oil *Environ Prog Sustain Energy* **33** pp. 1026-1033

[24] Othman N, Basri NEA, Yunus MNM, Sidek LM. 2008 Determination of physical and chemical characteristics of electronic plastic waste (Ep-Waste) resin using proximate and ultimate analysis method *International conference on construction and building technology* p. 169–80
[25] Rohit Kumar Singh, Biswajit Ruj, A.K. Sadhukhan, P. Gupta 2019 Impact of fast and slow pyrolysis on the degradation of mixed plastic waste: Product yield analysis and their characterization *Journal of the Energy Institute* Volume **92** Issue 6 Pages 1647-1657, ISSN 1743-9671 https://doi.org/10.1016/j.joei.2019.01.009.
[26] Chalita Kaewbuddee, Ekarong Sukjit, Jiraphon Srisertpol, Somkiat Maithomklang, Khatha Wathakit, Niti Klinkaew, Pansa Liplap and Weerachai Arjharn 2020 Evaluation of Waste Plastic Oil-Biodiesel Blends as Alternative Fuels for Diesel Engines *Energies* **13** 2823
[27] Ioannis Kalargaris, Guohong Tian, and Sai Gu 2016 Combustion, performance and emission analysis of a DI diesel engine using plastic pyrolysis oil *Fuel Processing Technology* vol. **157** https://doi:10.1016/j.fuproc.2016.11.016
[28] Uebe J, Kryzevicius Z, Majauskiene R, Dulevicius M, Kosychova L, Zukauskaite A 2022 Use of polypropylene pyrolysis oil in alternative fuel production *Waste Management & Research* https://doi:10.1177/0734242X211068243
[29] Viswanath K. Kaimal, P. Vijayabalan 2016 An investigation on the effects of using DEE additive in a DI diesel engine fuelled with waste plastic oil, Fuel, Volume 180, 2016, Pages 90-96, ISSN 0016-2361, https://doi.org/10.1016/j.fuel.2016.04.030.
[30] M. Parthasarathy, S. Ramkumar, P.V. Elumalai, Sachin Kumar Gupta, R. Krishnamoorthy, S. Mohammed Iqbal, Santosh Kumar Dash, R. Silambarasan 2021 Experimental investigation of strategies to enhance the homogeneous charge compression ignition engine characteristics powered by waste plastic oil *Energy Conversion and Management* Volume **236** 114026 ISSN 0196-8904 https://doi.org/10.1016/j.enconman.2021.114026.
[31] Manigandan Sekar, T.R. Praveenkumar, Veeman Dhinakaran, P. Gunasekar, Arivalagan Pugazhendhi 2021 Combustion and emission characteristics of diesel engine fueled with nanocatalyst and pyrolysis oil produced from the solid plastic waste using screw reactor Journal of Cleaner Production Volume 318 2021 128551 ISSN 0959-6526 https://doi.org/10.1016/j.jclepro.2021.128551.
[32] Viswanath K. Kaimal, P. Vijayabalan 2016 A study on synthesis of energy fuel from waste plastic and assessment of its potential as an alternative fuel for diesel engines *Waste Management* Volume **51** Pages 91-96 ISSN 0956-053X, https://doi.org/10.1016/j.wasman.2016.03.003.
[33] Viswanath K. Kaimal, P. Vijayabalan 2015 A detailed study of combustion characteristics of a DI diesel engine using waste plastic oil and its blends *Energy Conversion and Management* Volume **105** Pages 951-956 ISSN 0196-8904 https://doi.org/10.1016/j.enconman.2015.08.043.
[34] Thokchom Subhaschandra Singh, Tikendra Nath Verma, Huirem Neeranjan Singh 2020 A lab scale waste to energy conversion study for pyrolysis of plastic with and without catalyst: Engine emissions testing study *Fuel* Volume **277** 118176 ISSN 0016-2361 https://doi.org/10.1016/j.fuel.2020.118176.
[35] J. Devaraj, Y. Robinson, P. Ganapathi 2015 Experimental investigation of performance, emission and combustion characteristics of waste plastic pyrolysis oil blended with diethyl ether used as fuel for diesel engine Energy Volume 85 2015 Pages 304-309 ISSN 0360-5442 https://doi.org/10.1016/j.energy.2015.03.075.
[36] Sudershan B. Gadwal, N.R. Banapurmath, M.A. Kamoji, P.B. Rampure & S. V. Khandal 2019 Performance and emission characteristic studies on CRDI diesel engine fuelled with plastic pyrolysis oil blended with ethanol and diesel *International Journal of Sustainable Engineering* **12**:4, 262-271, DOI: 10.1080/19397038.2018.1527863
[37] J.B. Heywood 1984 Internal Combustion Engine Fundamentals *McGraw-Hill, USA*
[38] M. Mani, C. Subash, G. Nagarajan 2009 Performance, emission and combustion characteristics of a DI diesel engine using waste plastic oil *Appl. Therm. Eng.* **29** pp. 2738-2744
[39] S. Padmanabhan, C. Joel, Linda Joel, Obulareddy Yuvatejeswar Reddy, K.G.D. Sri Harsha and

S. Ganesan 2021 Evaluation of Waste Plastic Pyrolysis Oil Performance with Diethyl Ether Additive on Insulated Piston Diesel Engine *Nature Environment and Pollution Technology* Vol. **20** pp. 2079-2086 https://doi.org/10.46488/NEPT.2021.v20i05.025

[40] I. Ahmad, M.I. Khan, H. Khan, M. Ishaq, R. Tariq, K. Gul, et al. 2014 Pyrolysis study of polypropylene and polyethylene into premium oil products *Int J Green Energy* **12** pp. 663-671

[41] A. López, I. de Marco, B.M. Caballero, M.F. Laresgoiti, A. Adrados 2011 Dechlorination of fuels in pyrolysis of PVC containing plastic wastes *Fuel Process Technol* **92** pp. 253-260

[42] Amar Kumar Das, Dulari Hansdah, Achyut Kumar Panda 2021 Thermal balancing and exergetic performance evaluation of a compression ignition engine fuelled with waste plastic pyrolytic oil and different fuel additives *Energy* Volume **229** 120629 ISSN 0360-5442 https://doi.org/10.1016/j.energy.2021.120629.

[43] Miandad, R., Barakat, M. A., Aburiazaiza, A. S., Rehan, M., and Nizami, A. S. 2016 Catalytic pyrolysis of plastic waste: a review *Process Safety Environ. Protect.* **102** 822–838 doi: 10.1016/j.psep.2016.06.022

[44] Amankwa, M.O., Tetteh, E.K., Mohale, G.T. et al. 2021 The production of valuable products and fuel from plastic waste in Afric. *Discov Sustain* **2**, 31 https://doi.org/10.1007/s43621-021-00040-z

[45] Dedy Hariadi, Sofyan M. Saleh, R. Anwar Yamin, Sri Aprilia 2021 Utilization of LDPE plastic waste on the quality of pyrolysis oil as an asphalt solvent alternative *Thermal Science and Engineering Progress* Volume **23** 100872 ISSN 2451-9049 https://doi.org/10.1016/j.tsep.2021.100872.

[46] Direct emissions from recycling: 0.3–0.5 tonne CO2e per tonne of plastics recycled, and 1.6–3.3 tonnes CO2e per tonne of plastics produced from fossil-based virgin feedstock, depending on plastic resin type. (Deloitte, *Increased EU Plastics Recycling Targets: Environmental, Economic and Social Impact Assessment – Final Report*, 2015)

[47] Deng, C, Li, Y, Li, J, et al. 2017 Emission characteristics of PBDEs during flame-retardant plastics extruding process: field investigation and laboratorial simulation. *Environmental Science and Pollution Research* **24**: 22450–22457.

[48] Khan MZH, Sultana M, Al-Mamun MR, et al. 2016 Pyrolytic waste plastic oil and its diesel Blend: fuel characterization. *J Environ Public Health* 7869080.

[49] Eze WU, Madufor IC, Onyeagoro GN, et al. 2021 Study on the effect of Kankara zeolite-Y-based catalyst on the chemical properties of liquid fuel from mixed waste plastics (MWPs) pyrolysis. *Polym Bull* **78**: 377. doi: 10.1007/s00289-020-03116-4

[50] K. Naima and A. Liazid 2013 Review Waste oils as alternative fuel for diesel engine: A review *Journal of Petroleum Technology and Alternative Fuels* Vol. **4** (3)

[51] E. S. Wilkins & M. G. Wilkins 1985 Review of toxicity of gases emitted from combustion, pyrolysis of municipal and industrial wastes *Journal of Environmental Science and Health* Part A: Environmental Science and Engineering **20**:2 149-175, DOI: 10.1080/10934528509375217

Acknowledgments
The authors gratefully acknowledge the Ministry of Education (MOE) of Malaysia and Faculty of Ocean Engineering Technology and Informatics, University Malaysia Terengganu for the instruments under research grant Vot No. RACER/1/2019/WAB09/UMT//2 and UMT Vot: 59554.

AUTHOR INDEX

Institute of Physics
Dirac House, Temple Back
Bristol BS1 6BE UK

CURRAN ASSOCIATES INC.
proceedings.com

ISSN: 1755-1307
ISBN 978-1-7138-5455-5